Effects of Fatty Acids and Lipids in Health and Disease

World Review of Nutrition and Dietetics

Vol. 76

Basel · Freiburg · Paris · London · New York ·
New Delhi · Bangkok · Singapore · Tokyo · Sydney

Effects of Fatty Acids and Lipids in Health and Disease

Volume Editors   *Claudio Galli*
University of Milan

Artemis P. Simopoulos
The Center for Genetics, Nutrition and
Health, Washington, D.C.

Elena Tremoli
University of Milan

12 figures and 20 tables, 1994

 Basel · Freiburg · Paris · London · New York ·
New Delhi · Bangkok · Singapore · Tokyo · Sydney

....

World Review of Nutrition and Dietetics

Library of Congress Cataloging-in-Publication Data
Effects of fatty acids and lipids in health and disease / volume editors, Claudio Galli,
Artemis P. Simopoulos, Elena Tremoli.
(World review of nutrition and dietetics; vol. 76)
Includes bibliographical references and index.
1. Fatty acids – Physiological effect.   2. Fatty acids – Pathophysiology.   3. Fatty acids in human
nutrition.   4. Fatty Acids, Omega-3 – therapeutic use.   I. Galli, Claudio.   II. Simopoulos,
Artemis P., 1933–.   III. Tremoli, Elena.   IV. Series.
[DNLM: 1. Fatty Acids – metabolism.   2. Fatty Acids – therapeutic use.   3. Fatty Acids,
Omega-3 – metabolism. W1 WO898 v. 76 1994 / QU 85 F277 1994]
ISBN 3–8055–6040–0 (alk. paper)

Bibliographic Indices. This publication is listed in bibliographic services, including Current Contents® and
Index Medicus.

Drug Dosage. The authors and the publisher have exerted every effort to ensure that drug selection and
dosage set forth in this text are in accord with current recommendations and practice at the time of publication.
However, in view of ongoing research, changes in government regulations, and the constant flow of information
relating to drug therapy and drug reactions, the reader is urged to check the package insert for each drug for any
change in indications and dosage and for added warnings and precautions. This is particularly important when the
recommended agent is a new and/or infrequently employed drug.

© Copyright 1994 by S. Karger AG, P.O. Box, CH–4009 Basel (Switzerland)
Printed in Switzerland on acid-free paper by Thür AG Offsetdruck, Pratteln
ISBN 3–8055–6040–0

Contents

ω3 Fatty Acids and Thrombosis

Fatty Acids and Cancer

Inflammation and Immunology

Essential Fatty Acids, Pregnancy and Pregnancy Complications

Clinical Trials with ω3 Fatty Acids

International Society for the Study of Fatty Acids and Lipids

....
Preface

The Proceedings of the 1st International Congress of the International Society for the Study of Fatty Acids and Lipids (ISSFAL), held June 30–July 3, 1993, in Lugano, Switzerland, are presented in two volumes – volume 75 is entitled 'Fatty Acids and Lipids: Biological Aspects' and volume 76 is entitled 'Effects of Fatty Acids and Lipids in Health and Disease'.

First a brief background about ISSFAL. In March 1990, the 2nd International Conference on the Health Effects of ω3 Polyunsaturated Fatty Acids in Seafoods was held in Washington, D.C. (volume 66 in this series). It was recognized at that time that the field of research on ω3 polyunsaturated fatty acids (PUFA) had considerably advanced to the point that research on the role of *all* the fatty acids and lipids needed to expand. Therefore, it was decided that what was needed was one organization that would bring together scientists and clinical investigators interested in the role of fatty acids and lipids in health and disease. The name of the Society reflects exactly that, the International Society for the Study of Fatty Acids and Lipids – ISSFAL. The Organizing Committee for the Society consisted of Drs. Robert G. Ackman (Canada), Stuart Barlow (UK), Michael Crawford (UK), Jorn Dyerberg (Denmark), Claudio Galli (Italy), Robert Kifer (USA), William E.M. Lands (USA), Alexander Leaf (USA), Federico Leighton (Chile), Roy Martin (USA), Paul Nestel (Australia), Arne Nordøy (Norway), Rodolfo Paoletti (Italy), Serge Renaud (France), Artemis P. Simopoulos (USA), and Peter C. Weber (Germany). Dr. Leaf was appointed President, Dr. Dyerberg, Vice President, and Dr. Simopoulos, Secretary/Treasurer. In March 1991, ISSFAL was established as a nonprofit tax-exempt organization in the Commonwealth of Massachusetts (USA). The stated purpose of the Society is to increase understanding through research and education of the role of dietary fatty acids and lipids in health and disease.

The ISSFAL held its first congress on 'Fatty Acids and Lipids from Cell Biology to Human Disease' in Lugano, Switzerland, June 30–July 3, 1993. The Scientific Secretariat was headed by Drs. Claudio Galli and Elena Tremoli of the University of Milan, and the Organizing Secretariat was headed by Drs. Elena Columbo and Emanuela Folco of the Fondazione Giovanni Lorenzini in Milan, Italy. Four hundred and fifty persons from 35 countries attended, and a total of 332 abstracts consisting of state-of-the-art reviews, critiques and new data were presented at 13 plenary lectures, 10 major symposia, 10 oral communication sessions, and 14 poster sessions, covering topics such as Intracellular Communication; A New Look at Fatty Acids as Signal Transducing Molecules; ω3 PUFA in the Regulation of Cytokine Synthesis; The Role of Fatty Acids during Pregnancy and Lactation; Fatty Acids and Human Physiology; Cardiovascular System; Hypertension; Diabetes; Cancer; Inflammation and Immunology; PUFA and Antioxidants; a round table discussion on The Future of Fatty Acids in Human Nutrition, Health and Policy Implications, and a final session in which summary statements were presented by the session chairmen for general discussion. During this final session it was recommended that trans fatty acids should be labelled as such and should not be included with other fatty acids, such as saturates or PUFA. The need for standardization of fatty acid and lipid nomenclature was recognized, and at the Board Meeting of ISSFAL, a subcommittee was appointed by Dr. Leaf to develop a position paper.

These Proceedings consist of two volumes. Volume 75 includes a summary of the round table discussion on the Future of Fatty Acids in Human Nutrition: Health and Policy Implications; and the Health Message Statement on fatty acids developed by a subcommittee of the ISSFAL Board and approved by a majority of the Board members. Following are the papers presented at the sessions on Enzymes of PUFA Metabolism and Oxidation; Fatty Acid Oxidation; Fatty Acids and Cell Signalling; Fatty Acids and Human Physiology; Maternal and Infant Nutrition; Mechanisms of Accretion of Polyunsaturates in the Nervous System; PUFA and Natural Antioxidants, and Isomeric Fatty Acids.

Volume 76 includes the papers presented at the sessions on Fatty Acids and Cardiovascular System; ω3 and ω6 Fatty Acids, Lipids and Lipoproteins; ω3 Fatty Acids and Thrombosis; Fatty Acids and Cancer; Inflammation and Immunology; Essential Fatty Acids; Pregnancy and Pregnancy Complications, and Clinical Trials with ω3 Fatty Acids.

These Proceedings and the congress abstracts contain the most up-to-date information on the physiological and metabolic aspects of fatty acids and lipids in health and disease, in the form of reviews, new data, and state-of-the-art papers on the role of ω3 fatty acids and their relationship to ω6, ω9, saturated fats and trans fatty acids in biomedical research and in clinical investigations. The Proceedings should be of interest to biomedical researchers in academia, industry and government, including clinical investigators, physiologists, biochemists, nutritionists, dietitians, and policy-makers.

Artemis P. Simopoulos, MD
Claudio Galli, MD
Elena Tremoli, PhD

Galli C, Simopoulos AP, Tremoli E (eds): Effects of Fatty Acids and Lipids in
Health and Disease. World Rev Nutr Diet. Basel, Karger, 1994, vol 76, pp 1–8

....

Some Effects of ω3 Fatty Acids on Coronary Heart Disease[1]

Alexander Leaf

Harvard Medical School and Massachusetts General Hospital, Boston, Mass. and
West Roxbury VA Medical Center, West Roxbury, Mass., USA

Although there had been a few reports of beneficial effects of eating fish on
heart disease [1, 2], prior to the seminal epidemiologic studies of Greenland
Inuits by Bang et al. [3], it was clearly the latter that initiated the present and
growing interest in a possible role of ω3 fatty acids in the prevention of
cardiovascular disease, especially of atherosclerosis and its chief clinical mani-
festation, coronary heart disease (CHD). Dyerberg and Bang surmised that the
low incidence of mortality from CHD among the Eskimos, despite a high total
fat diet, might be attributable to the relatively large intake of ω3 fatty acids
from the high proportion of marine vertebrates in their diets. Today that
suggestion seems to be correct, as every effect of fish and marine mammals that
has been tested on factors involved in CHD, seems replicable by two long chain
polyunsaturated fatty acids, eicosapentaenoic acid (c20:5 ω3, EPA) and docosahexaenoic acid (c22:6 ω3, DHA) which are present in oils of fish and marine
mammals. Use of the term ω3 fatty acids here will refer to these two long chain
highly polyunsaturated fatty acids and not, unless specifically stated, to α-
linolenic acid (c18:3 ω3, LNA) the parent essential fatty acid of this ω3 series
which is present in some vegetable oils, because of uncertainty that the
elongation and desaturation rates of LNA are sufficient to provide all the needs
of humans for EPA and DHA.

As might be expected, the search for the action of ω3 fatty acids which
could account for their protection against CHD has followed the advances in
our understanding of the pathogenesis of atherosclerosis. For this reason there

[1] The studies on the effects of fish oil fatty acids on calcium channels were supported by
Public Health Service, NIH grants RO1 DK38165 and UO1 HL40548.

have been a surfeit of studies to examine the effects of ω3 fatty acids on blood lipid levels. By 1989, when Harris [4] reviewed the literature on this topic, there were already over 40 reported studies. Since then the number has probably doubled despite the fact that only a dramatic lowering of triglyceride levels is produced by ingestion of fish oils. A reduction of total and low density lipoprotein (LDL) cholesterol occur only when ω3 fatty acids replace a high intake of saturated fatty acids in the diet, an effect shared by ω6 fatty acids. Otherwise, the effects on total and LDL cholesterol are essentially nil. The effects on high density lipoprotein (HDL) cholesterol are also minimal and variable. An effect of ω3 fatty acids to produce a physically distinct form of LDL particles which are less atherogenic has been suggested but requires further study.

Unfortunately there are many firm believers that cholesterol levels are the 'be all and end all' of CHD and no substance that does not significantly improve the blood lipid profile is worthy of further investigations. This view seems to be especially dear to those who fund medical research despite the convincing evidence that aspirin, which has no effect on plasma cholesterol levels, has been found to reduce the incidence of CHD. Fortunately, there are others who take a much broader view of the causes of atherosclerosis and who have extended the search for ways that fatty acids may affect the process. Much progress has resulted.

Atherosclerosis is now thought to be initiated by a dysfunctional state of the arterial endothelial cells. This may result from many stresses including hemodynamic, elevated LDL cholesterol, virus infections, toxins, trauma, etc. The first visible effect of dysfunction is the adhesion of circulating monocytes to the affected endothelial cells. This is soon followed by penetration of the endothelium into the underlying intima by these cells where they remain as resident macrophages. Release of oxygen free radicals by macrophages and endothelial cells oxidize the lipids and alter the apoprotein of the LDL lipoproteins which normally shuttle between plasma and intima. The oxidized LDL particles are then recognized by the scavenger receptors and taken up by the macrophages to commence lipid accumulation in the incipient atherosclerotic plaque. Dysfunctional endothelial cells and macrophages are soon joined by migrating and proliferating smooth muscle cells and fibroblasts and by circulating neutrophiles, lymphocytes, and platelets which produce a concert of growth factors, chemoattractants, cytokines, leukotrienes, prostaglandins and other cellular messengers which, together with episodic thrombotic events, contribute to the progression of the atherosclerotic changes. In fact it is now appreciated that virtually the full armamentarium of the body's thrombotic, proliferative, inflammatory and immunological activities contribute to the atherosclerotic process (fig. 1). This means that very many complex activities

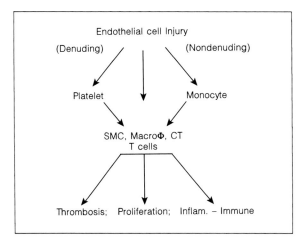

Fig. 1. The relationship of thrombotic, proliferative, inflammatory and immunologic processes to the pathogenesis of atherosclerosis. These major defense mechanisms of our bodies are all mobilized in the pathogenesis of atherosclerosis.

which are genetically controlled participate in the pathogenesis of atherosclerosis. A corollary of this is that many genetically controlled factors which in an individual respond excessively or minimally to environmental risk factors make a person either more or less vulnerable to developing atherosclerosis in response to a given atherosclerotic stress.

ω3 fatty acids, where tested, affect a number of these factors – see table 1, which is an incomplete listing. Other physiologic and pharmacologic actions of ω3 fatty acids are listed in table 2. What to me seems so remarkable about these fatty acids is that those factors which have atherogenic potential are downregulated by ω3 fatty acids whereas those that have antiatherogenic potential, such as prostacyclin and endothelium-derived relaxation factor (EDRF), are enhanced by the presence of ω3 fatty acids.

How can a naturally occurring dietary factor possess so many different effects in our bodies? When I first heard the multiple claims made for these fatty acids, I thought they represented another example of the mythical claims for 'snake oil'. I am now convinced, however, that many of the claims are possible and that perhaps the new discipline of evolutionary medicine can provide an explanation for these multiple effects. Medical anthropologists have studied the human diet of our forebears and have concluded that during some 2–4 million years our ancestors had existed primarily as hunter-gatherers [5]. By analysis of the diets of the few remaining hunter-gatherer cultures, they have determined that during the long period of human development, during which our genetic

Table 1. Factors affecting atherogenesis

Factor	Effect of ω3 fatty acid	Function
1. Thromboxane, TXA$_2$ [15]	↓	platelet aggregation vasoconstriction
2. Prostacyclin, PGI [16]	↑	prevents platelet aggregation vasodilation
3. Leukotriene, LTB$_4$ [17]	↓	neutrophil chemoattractant and aggregator
4. Platelet-activating factor, PAF [18]	↓	activates platelets
5. Platelet-derived growth factor, PDGF [19]	↓	chemoattractant and mitogen for smooth muscles and macrophages
6. Superoxide formed by leukocytes [20]	↓	cellular damage; enhances LDL uptake by macrophages
7. Interleukin-1, IL-1, and tumor necrosis factor, TNF [21]	↓	expresses endothelial adhesion molecule-1 stimulates PAF; inhibits plasminogen activator; stimulates smooth muscle cell proliferation; stimulates neutrophil superoxide formation
8. Endothelial-derived relaxation factor, EDRF [22, 23]	↑	reduces arterial constrictor responses protects endothelial surface from thrombi

Table 2. Additional physiologic and pharmacologic effects of fish oils

1. Decreases blood pressure in normal and moderately hypertensive subjects [24]
2. Decreases blood viscosity [25]
3. Decreases microvascular albumin leakage in insulin-dependent diabetics [26]
4. Decreases plasma triglycerides [27]
5. Decreases vascular response to norepinephrine [28]
6. Decreases ventricular fibrillation from ischemia [9]
7. Decreases cardiac toxicity of cardiac glycosides in vitro [15]
8. Decreases platelet adhesion [29]
9. Decreases leukocyte/endothelial interactions [30]
10. Increases vascular compliance [31]
11. Increases thrombolytic activity of TPA [32]
12. Increases platelet survival [33]

Leaf

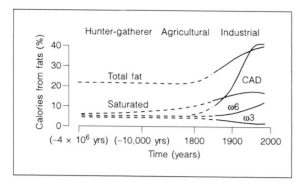

Fig. 2. Hypothetical scheme of the relative percentages of fat and of different fatty acid families in human nutrition as extrapolated from cross-sectional analyses of contemporary hunter-gatherer populations and from longitudinal observations and their putative changes during the past 200 years of the Industrial Revolution in relation to the recent increase in the frequency of CHD [from 6, with permission.]

makeup was established, the diet of our forebears contained fatty acids and lipids approximately in the proportions depicted by Leaf and Weber [6] in figure 2. The total fat in the diet contributed only slightly more than 20% of energy and ω3 fatty acids were abundant and quantitatively only slightly less than were ω6 fatty acids, the other class of essential fatty acids. Their source was not only from the α-linolenic acid in plants and vegetables, but mainly preformed from the meat of range fed animals. It was only recently, some 10–15 thousand years ago, when our ancestors discovered agriculture and animal husbandry that diets commenced to change. This has been greatly accelerated since the start of the Industrial Revolution. Cattle were fattened, vegetable oils hydrogenated and foods processed. The total, saturated, and trans fatty acids in diets increased, as did the ω6 fatty acids while the ω3 fatty acids diminished nearly to vanish from the diets in industrialized Western countries. It is with these recent changes in lipid intake that the epidemic of CHD has occurred. Today, where these diets spread to other countries, so does the accompanying scourge of CHD morbidity and mortality. It is now well recognized that CHD is a disease of affluent societies ingesting a surfeit of total and saturated fats, but I would suggest that it is also partially a deficiency disease due to a lack of adequate ω3 fatty acids in the Western diet. During the long evolutionary period of exposure to ω3 fatty acids they developed many important physiologic and biochemical functions in our bodies just as has arachidonic acid (c20:4 ω6, AA) which has many well-recognized actions and regulatory roles.

There have now been many studies of the effectiveness of the ω3 fatty acids to prevent experimental atherosclerosis in animals. Most have demonstrated a protective effect, but some have failed to do so. Human studies are still few in number. The initial observation in the Zutphen study by Kromhout et al. [7] of an inverse relation between fish ingestion and mortality from CHD has received strong support from the recent analyses of the large Multiple Risk Factor Intervention Trial by Dolcek [8]. She divided the 6,000 subjects in the control group (Usual Care) for that Trial into quintiles according to their mean ingestion of ω3 fatty acids from 0 to 0.66 g daily and found significant inverse correlations between the ingestion of these fish oil fatty acids and CHD, all cardiovascular diseases and all causes of mortality.

Another effect of ω3 fatty acids on CHD has been recently investigated by McLennan et al. [9]. They have demonstrated that a diet rich in tuna fish oil essentially prevents fatal ventricular fibrillation following experimental myocardial infarction in rats. They also found the threshold for induction of ventricular fibrillation by electrical stimulation to be increased in nonhuman primates by a diet high in ω3 fatty acids [10]. Following these pioneering findings, Hallaq et al. [11, 12] in my laboratory have shown that EPA, or, even more effectively, DHA can prevent the toxic arrhythmias induced in isolated rat heart cells by ouabain. The mechanism of this preventive effect is by modulation of L-type calcium channels in the sarcolemma of the cardiac myocytes in a manner that prevents the cytosolic free calcium concentrations from reaching toxic levels. But intriguingly the ω3 fatty acids are not acting simply as calcium channel blockers. Rather they are able to reduce calcium influx into the myocytes when levels of free calcium in the cytosol threaten contracture and fibrillation of the cells or to increase calcium influx into the myocytes in response to dihydropyridine calcium channel antagonists, e.g. nitrendipine, when the latter so reduce calcium influx via L-type calcium channels that the myocytes no longer contract. In either case they protect cytosolic calcium levels so as to sustain normal contractility of the myocytes.

The potential significance of this antiarrhythmic action of ω3 fatty acids is accentuated by the findings of Burr et al. [13] in the DART study. In that secondary prevention trial, advice to eat fish or take a fish oil supplement significantly reduced CHD mortality compared with a randomized control group, but did not reduce the incidence of new cardiac events. If one had a second heart attack, he was less likely to die from it. Since some 60% of the mortality from acute myocardial infarctions results from sudden death [14] due to ventricular fibrillation (some 300,000 of the 500,000 death annually in the US), this action of ω3 fatty acids suggests they protect against fatal ventricular fibrillation and may thus have considerable public health benefit.

Because the ω3 fatty acids, or their partially oxidized metabolites, may modify so many different factors which can affect the progression or prevention of atherosclerosis, it seems a vain hope that a single action of these fatty acids can be identified, to which all the beneficial effects can be attributed. Much effort has been expended to demonstrate that the antiplatelet action of the ω3 fatty acids compares unfavorably with that of even low doses of aspirin. But that is indeed fortunate. Were a normal dietary constituent to abolish, rather than diminish, the thrombotic action of platelets and the diverse effects of the many other factors affected by the ω3 fatty acids, it could put us at risk of serious consequences. But fortunately their actions only modulate, rather than abolish, the actions of these vital factors.

The effects of the ω3 fatty acids are physiologic, not pharmacologic or pathologic. A corollary to this fact is that many of their actions will be small and subtle, thus increasing the difficulty of experimental detection. Without the application of great care and excellent research techniques, perhaps in some studies the effects may be lost in the experimental error. A further caveat is that the ω3 fatty acids should not be expected to prevent all the causes of atherosclerosis. When the atherogenic stimuli are very potent they may over-whelm the beneficial actions of these dietary ingredients. But either ingested regularly as fish or as a supplement to a prudent, low saturated fat and cholesterol diet, it is becoming increasingly evident that the ω3 fatty acids may help to reduce the burden of coronary heart disease in the world.

References

1   Nelson AM: Diet therapy in coronary heart disease – Effect on mortality of high-protein, high-seafood, fat-controlled diet. Geriatrics 1972;27:103–116.
2   Pfeifer JJ: Hypocholesterolemic effects of marine oils; in Stansby ME (ed): Fish Oils. Westport, Avi Publishing Co, 1967, chapt 23.
3   Bang HO, Dyerberg J, Horne N: The composition of food consumed by Greenland Eskimos. Acta Med Scand 1976;200:69–73.
4   Harris WS: Fish oils and plasma lipid and lipoprotein metabolism in humans; a critical review. J Lipid Res 1989;30:785–807.
5   Eaton SB, Konnor M: Paleolithic nutrition: A consideration of its nature and current implications. N Engl J Med 1985;312:283–289.
6   Leaf A, Weber PC: A new era for science in nutrition. Am J Clin Nutr 1987;45:1048–1053.
7   Kromhout D, Bosschieter EB, de Lezenne Coulander C: The inverse relation between fish consumption and 20-year mortality from coronary heart disease. N Engl J Med 1985;312:1205–1209.
8   Dolcek TA: Epidemiological evidence of relationships between dietary polyunsaturated fatty acids and mortality in the Multiple Risk Factor Intervention Trial. PSEBM 1992;200:177–182.
9   McLennan PL, Abeywardena MY, Charnock JS: The influence of age and dietary fat in an animal model of sudden cardiac death. Aust NZ J Med 1989;19:1–5.
10 McLennan PL, Bridle TM, Abeywardena MY, Charnock JS: Dietary lipid modulation of ventricular fibrillation threshold in the marmoset monkey. Am Heart J 1992;123:1555–1561.
11 Hallaq H, Sellmayer A, Smith TW, Leaf A: Protective effect of eicosapentaenoic acid on ouabain toxicity in neonatal rat cardiac myocytes. Proc Natl Acad Sci USA 1990;87:7834–7838.
12 Hallaq H, Smith TW, Leaf A: Modulation of dihydropyridine-sensitive calcium channels in heart cells by fish oil fatty acids. Proc Natl Acad Sci USA 1992;89:1760–1764.
13 Burr ML, Gilbert JF, Holliday RM, Elwood PC: Effects of changes in fat, fish and fibre intakes on death and reinfarction: Diet and reinfarction trial (DART). Lancet 1989;ii:757–761.

14 American Heart Association (AHA): 1990 Heart and Stroke Facts. Dallas, AHA, 1989.

15 Dyerberg J, Bang HO, Stofferson E, Moncada S, Vane JR: Eicosapentaenoic acid and prevention of thrombosis and atherosclerosis? Lancet 1978;ii:117–119.

16 Von Schacky C, Fischer S, Weber PC: Long-term effect of dietary marine omega-3 fatty acids upon plasma and cellular lipids, platelet function, and eicosanoid formation in humans. J Clin Invest 1985;76:1626–1631.

17 Lee TH, Hoover RL, Williams JD, et al: Effects of dietary enrichment with eicosapentaenoic and docosahexaenoic acids on in vitro neutrophil and monocyte leukotriene generation and neutrophil function. N Engl J Med 1985;312:1217–1224.

18 Sperling RI, Robin JL, Kylander KA, Lee TH, Lewis RA, Austen KF: The effects of ω3 polyunsaturated fatty acids on the generation of platelet-activating factor-acether by human monocytes. J Immunol 1987;139:4186–4191.

19 Fox PL, Di Corleto PE: Fish oils inhibit endothelial cell production of platelet-derived growth factor-like protein. Science 1988;241:453–456.

20 Fisher M, Levine PH, Weiner BH, Johnson MH, Doyle EM, Ellis PA, et al: Dietary ω3 fatty acid supplementation reduces superoxide production and chemiluminescence in a monocyte-enriched preparation of leukocytes. Am J Clin Nutr 1990;51:804–808.

21 Endres S, Ghorbanc R, Kelley VE, Georgills K, Konnemanms G, et al: The effect of dietary supplementation with ω3 polyunsaturated fatty acids on the synthesis of interleukin-1 and tumor necrosis factor by mononuclear cells. N Engl J Med 1989;320:265–271.

22 Shimokawa H, Vanhoutte PM: Dietary ω3 fatty acids and endothelium-dependent relaxations in porcine coronary arteries. Am J Physiol 1989;256:H968–H973.

23 Malis C, Leaf A, Varadarajan GS, Newell JB, Weber PC, Force T, Bonventre JV: Effects of dietary fish oil on vascular contractility in preanoxic and postanoxic aortic rings. Circulation 1991; 84:1393–1401.

24 Knapp HR, FitzGerald GA: The antihypertensive effects of fish oil. N Engl J Med 1989;320:1037–1043.

25 Terano T, Hirai A, Hamazaki T, et al: Effect of oral administration of highly purified eicosapentaenoic acid on platelet function, blood viscosity and red cell deformability in healthy human subjects. Atherosclerosis 1983;46:321–331.

26 Jensen T, Stender S, Goldstein K, Holmer G, Deckert T: Partial normalization by dietary cod liver oil of increased microvascular albumin leakage in patients with insulin-dependent diabetes and albuminuria. N Engl J Med 1989;321:1572–1577.

27 Phillipson BE, Rothrock DW, Connor WE, Harris WS, Illingsworth DR: Reduction of plasma lipids, lipoproteins and apolipoproteins by dietary fish oil in patients with hypertriglyceridemia. N Engl J Med 1985;312:1210–1216.

28 Lorenz R, Spengler U, Fischer S, Duhm J, Weber PC: Platelet function, thromboxane formation and blood pressure control during supplementation of the Western diet with cod liver oil. Circulation 1983;67:504–511.

29 Li X, Steiner M: Fish oil: A potent inhibitor of platelet adhesiveness. Blood 1990;76:938–945.

30 Lehr H-A, Hubner C, Finckh B, Noite D, Beisiegel U, Kohlschutter A, Messmer K: Dietary fish oil reduces leukocyte/endothelium interaction following systemic administration of oxidatively modified low density lipoprotein. Circulation 1991;84:1725–1731.

31 Wahlquist ML, Lo CS, Myers KA: Fish intake and arterial wall characteristics in healthy people and diabetic patients. Lancet 1989;ii:944–946.

32 Braden GA, Knapp HR, Fitzgerald DJ, Fitzgerald GA: Dietary fish oil accelerates the response to coronary thrombolysis with tissue plasminogen activator: Evidence for a modest platelet inhibitory effect in vivo. Circulation 1990;82:178–187.

33 Levine PH, Fisher M, Schneider PB, Whitten RH, Weiner BH, et al: Dietary supplementation with omega-3 fatty acids prolongs platelet survival in hyperlipidemic patients with atherosclerosis. Arch Intern Med 1989;149:1113–1116.

Alexander Leaf, MD, West Roxbury VA Medical Center, 1400 VFW Parkway,
West Roxbury, MA 02132 (USA)

Galli C, Simopoulos AP, Tremoli E (eds): Effects of Fatty Acids and Lipids in
Health and Disease. World Rev Nutr Diet. Basel, Karger, 1994, vol 76, pp 9–14

....

Fatty Acids and Hypertension

Howard R. Knapp

Division of Clinical Pharmacology, Department of Internal Medicine,
University of Iowa, Iowa City, Iowa, USA

The influence of dietary fat on blood pressure is complex, and it is likely that both the amount and type of fats consumed exert different effects on vascular control mechanisms. This review will focus upon dietary polyunsaturated fatty acids as modulators of blood pressure. The epidemiology and clinical trial experience with ω6 polyunsaturates is largely limited to those providing dietary 18-carbon compounds, as these are the prevalent polyunsaturates in the major grain crops in the US and Europe. This subject was reviewed [1], and experimental evidence has revealed that, under a number of circumstances, additional intake of ω6 polyunsaturates has little effect on blood pressure in individuals who already consume a diet rich in these fatty acids. This presentation, therefore, will deal largely with dietary enrichment with the long-chain ω3 fatty acids found in marine oils, and because of space limitations reference will sometimes be made to previous review articles rather than to the extensive primary literature.

Epidemiology of Dietary ω3 Fatty Acids and Blood Pressure

Although some early reports described low blood pressure in hospitalized Canadian and Greenland Eskimos, subsequent surveys of large numbers of free-living subjects in Alaska [2] or Greenland [3, 4] found no difference in blood pressure or its age-related increase between Eskimos and Caucasians. The epidemiology of dietary ω3 fatty acids and blood pressure is confounded by the fact that populations consuming large amounts of fish or marine mammals also have a relatively high intake of sodium. If one compares blood pressures in different regions of the same country or in migration studies, it is apparent that blood pressure tracks with the intake of sodium rather than that of ω3 fatty acids [5]. Experimental evidence has also indicated that a high intake of sodium can obscure any blood pressure-lowering effect of ω3 fatty acids, while sodium restriction can accentuate it [6]. Thus, it is not possible to find clear evidence for a blood pressure-lowering effect of ω3 fatty acid ingestion from dietary surveys of populations.

Clinical Studies of ω3 Fatty Acid Vascular Effects

There have been reports that dietary enrichment with either ω3 or ω6 fatty acids lowers vascular reactivity, platelet aggregation, and blood pressure [reviewed in 7], and an additional problem in understanding the vascular effects of polyunsaturated fatty acids is the failure to compare both classes of compounds in the same studies. In this way, it would be possible to determine the specificity of the physiological effects found, but this has been done only occasionally. Also, since eicosanoids are made from polyunsaturated fatty acids and exert a number of potent influences on blood pressure control, the effects of dietary polyunsaturate manipulation are frequently ascribed to altered eicosanoid synthesis, but this has rarely been directly assessed in diet-blood pressure studies. We have discussed the important aspects of study design that are frequently omitted in nutritional studies of hypertension [5], and published the first study utilizing ambulatory blood pressure monitors and GC/MS measurement of urinary eicosanoid metabolites in patients with mild essential hypertension [8]. Both high (15 g/day) and low (3 g/day) doses of ω3 fatty acids as menhaden oil were compared with a high dose of linoleic acid (35 g/day) given as safflower oil and a control oil mixture mimicking the fatty acid composition of the average American diet. Only the high-dose fish oil group had a statistically significant decline in both systolic and diastolic blood pressures. Changes in the urinary metabolites of thromboxanes and prostaglandins of the E-type did not explain the blood pressure fall, but a marked increase in prostacyclin metabolite was seen early on, and we hypothesized that this might have resulted in alterations in other blood pressure-regulating systems. Our findings were subsequently confirmed and extended in a large population-based study in Norway, in which hypertensive patients who did not habitually eat fish had a reduction in blood pressure with 6 g eicosapentaenoic acid/day but not with a similar amount of corn oil [9]. These two studies by now join several others in providing evidence for an antihypertensive effect of dietary ω3 fatty acid supplements [10], but the possible mechanism of this effect remains unknown.

One type of hypertension which is associated with excess thromboxane synthesis and which has been ameliorated in a number of cases with platelet-selective doses of aspirin is that induced by pregnancy [11]. Although we did not find that dietary fish oil caused a marked reduction in the normal levels of thromboxane being produced by our healthy males with mild essential hypertension [8], it is conceivable that such a benefit could occur in women making abnormally elevated levels, as we saw in our earlier studies in severe peripheral vascular disease [12]. It is of interest, therefore, that lower, more palatable doses of ω3 fatty acids may be found to exert beneficial effects on blood pressure when taken for long periods of time, as suggested by the recent reanalysis of a study in England from 1938–39 [13]. Pregnant women participated in a randomized trial of nutritional supplements that included a low dose of herring oil providing <0.5 g/day eicosapentaenoic acid. It was found that the number of premature births, the rate of proteinuria, and the incidence of both pregnancy-induced and pregnancy-exacerbated hypertension was lowered to a highly significant degree in the women receiving the supplement.

Possible Mechanisms of ω3 Fatty Acid Effects

The antihypertensive effect of ω3 fatty acids could be via several different mechanisms in conditions having varying etiologies, and suggestions have been made for a number of possibilities. These include, among others, reduction in blood viscosity [14] which by itself would lower blood pressure, reduced response to endogenous pressors [15], and altered baroreceptor function [10]. The first study reporting a reduced pressor response to norepinephrine was that of Lorenz et al. [15], but this was intended to be a biochemical study, and the subjects were not on controlled diets or studied in a randomized way. The subjects had an unintentional increase in salt intake during the study, but surprisingly had a nonsignificant decline in angiotensin responsiveness rather than the increase that one would expect in this situation. Subsequent efforts to detect an effect of ω3 fatty acid supplements on catecholamine response have been unsuccessful, with two groups finding no difference to endogenous catecholamines during stress testing before and during supplementation [16, 17]. On the other hand, it has been recently shown that the vascular response to angiotensin, but not norepinephrine, in the human forearm is blunted by dietary ω3 fatty acids [18]. An earlier study had also shown reduced pressor response to systemic angiotensin infusion after ω3 fatty acid supplementation [19].

We attempted to address the issue of systemic catecholamine response by performing phenylephrine infusions in mildly hypertensive patients at the end of 4 weeks of menhaden oil supplements or their 4-week posttreatment period [20]. There was no difference in the blood pressure response to moderate phenylephrine infusion rates, but we made the unexpected observation that the patients developed a reflex bradycardia at a lower blood pressure while they were taking fish oil. Interestingly, this difference was not seen in the patients taking the control fatty acid mixture, but was seen in patients taking both the high-dose and low-dose fish oil supplements; only the former of these was associated with a lowering of blood pressure. Other workers have observed a similar phenomenon [21] in that normotensive subjects in whom no reduction in blood pressure was detected had up-regulation of baroreflex function to a norepinephrine infusion when taking fish oil supplements. This finding, and our seeing enhanced baroreflex activity in our low-dose ω3 group, indicates that the enhanced baroreflex is not due to 'resetting' as a result of blood pressure reduction [22], but rather may be a primary effect that could eventually lead to a reduction in abnormally elevated blood pressure.

Since prostacyclin infusion has been found to stimulate baroreflex function [23], we assessed its synthesis and possible involvement in the baroreflex alterations in our patients by measuring the excretion of its main urinary metabolite by GC/MS [8]. Pressor infusions have been found to cause modest increases in prostacyclin production in normal subjects, presumably due to increased vascular wall mechanical stress. We also found this to occur in our three groups of subjects when they were off their supplements, and in the control oil mixture group during supplementation. Both fish oil groups, however, had higher baseline production prior to their phenylephrine infusions, and an actual reduction in prostacyclin during the infusion rather than an increase. It is unknown whether the fact that their pressures did not achieve as high a level on this occasion, due to bradycardia, accounts for this finding. Clearly, the possible importance of chronic baroreflex control in blood pressure regulation [22] is being studied intently, as well as its apparent up-regulation by dietary ω3 fatty acids.

In addition to alterations in baroreflex control and pressor responsiveness, it is possible that dietary ω3 fatty acids alter other aspects of blood pressure regulation. The question of lowering blood viscosity is an important one that has been reviewed previously [24], but contradictory results continue to appear in the literature. A blinded comparison of ω3 and ω6 fatty acid effects on erythrocyte flexibility, with repeated measures and a recovery period, would be a very worthwhile addition to sorting out whether any of the effects of ω3 fatty acids could be via this simple mechanical mechanism. An interesting observation in a rat hypobaric hypoxia model [25] suggested that animals given dietary ω3 fatty acids had improved mortality, lower pulmonary artery pressures and less right ventricular hypertrophy, possibly on the basis of reduced blood viscosity (with hematocrits being the same). Thus, the issue of ω3 fatty acid effects on blood viscosity could impact on possible benefits in areas other than systemic hypertension and via mechanisms not involving interactions with the eicosanoid system.

The recent development of drugs that effectively inhibit the 5-lipoxygenase has led to clinical studies [26] that have not revealed a significant effect on blood pressure or heart rate. Thus, it appears unlikely that the vascular effects of ω3 fatty acids could be via alterations in leukotriene synthesis. Another noncyclooxygenase family of compounds with reported vasoactive properties are the 8-epi-PGF compounds described by Morrow et al. [27] as being autooxidation products of arachidonate that may interact with thromboxane receptors. Virtually nothing is known of how dietary ω3 fatty acids alter the types or amounts of these compounds produced in either normal or hypertensive humans, and this is an important area for future work. The epoxides and diols derived by cytochrome P-450 oxygenation of arachidonic acid likewise are being intensively studied for their possible involvement in human vascular disease. We have reported that analogous compounds are also generated from eicosapentaenoic acid by human subjects ingesting marine oils [28], and that markedly increased amounts are produced by hypertensive subjects as their blood pressure falls during ω3 fatty acid supplementation [29]. Understanding the role of these various fatty acid oxygenation products in mediating the effects of ω3 fatty acids may help us to both define the control mechanisms involved in human vascular disease as well as determine the types of vascular patients that would be most likely to benefit from therapy with ω3 fatty acids.

References

1   Sacks FM: Dietary fats and blood pressure: A critical review of the evidence. Nutr Rev 1989;47: 291–300.
2   Scott EM, Griffith IV, Hoskins DD, et al: Serum-cholesterol levels and blood pressure of Alaskan Eskimo men. Lancet 1958;88:667–668.
3   Ehrstrom I: Medical studies in North Greenland 1948–1949. Acta Med Scand 1951;40:416–422.
4   Bjerager P, Kromann N, Thygesen K, et al: Blodtryk hos Gronlaendere. Ugeskr Læger 1980;142: 2278–2280.
5   Knapp HR: Omega-3 fatty acids, endogenous prostaglandins, and blood pressure regulation in humans. Nutr Rev 1989;47:301–313.
6   Cobiac L, Nestel PJ, Wing LMH, et al: Effects of dietary sodium restriction and fish oil supplements on blood pressure in the elderly. Clin Exp Pharmacol Physiol 1991;18:265–268.
7   Knapp HR, Whittemore KL, FitzGerald GA: Dietary polyunsaturates and human vascular function; in Lands WEM (ed): Proceedings of the AOCS Short Course on Polyunsaturated Fatty Acids and Eicosanoids. Champaign, American Oil Chemists' Society, 1987, pp 41–55.
8   Knapp HR, FitzGerald GA: The antihypertensive effects of fish oil: A controlled study of polyunsaturated fatty acid supplements in essential hypertension. N Engl J Med 1989;320:1037–1043.
9   Bønaa KH, Bjerve KS, Straume B, et al: Effect of eicosapentaenoic and docosahexaenoic acids on blood pressure in hypertension: A population-based intervention trial from the Tromsø study. N Engl J Med 1990;322:795–801.
10   Knapp HR: n-3 fatty acids and blood pressure regulation in man; in Frolich JC, von Schacky C (eds): Fish, Fish Oil and Human Health, Clinical Pharmacology. Munich, Zuckschwerdt, 1993, vol 5, pp 112–122.

11 Fitzgerald DJ, FitzGerald GA: Eicosanoids in the pathogenesis of pre-eclampsia; in Laragh JH, Brenner BM (eds): Hypertension: Pathophysiology, Diagnosis, and Management. New York, Raven Press, 1989.
12 Knapp HR, Reilly IAG, Alessandrini P, et al: In vivo indexes of platelet and vascular function during fish-oil administration in patients with atherosclerosis. N Engl J Med 1986;314:937–942.
13 Olsen SF, Secher NJ: A possible preventive effect of low-dose fish oil on early delivery and pre-eclampsia: Indications from a 50-year-old controlled trial. Br J Nutr 1990;64:599–609.
14 McMillan DE: Antihypertensive effects of fish oil. N Engl J Med 1989;321:1610.
15 Lorenz R, Spengler U, Fischer S, et al: Platelet function, thromboxane formation and blood pressure control during supplementation of the Western diet with cod liver oil. Circulation 1983; 67:504–511.
16 Singer P, Wirth M, Voigt S, et al: Blood pressure- and lipid-lowering effect of mackerel and herring diet in patients with mild essential hypertension. Atherosclerosis 1985;56:223–235.
17 Hughes GS Jr, Ringer TV, Francom SF, et al: Effects of fish oil and endorphins on the cold pressor test in hypertension. Clin Pharmacol Ther 1991;50:538–546.
18 Kenny D, Warltier DC, Pleuss JA, et al: Effect of omega-3 fatty acids on the vascular response to angiotensin in normotensive men. Am J Cardiol 1992;70:1347–1352.
19 Yoshimura T, Matsui K, Ito M, et al: Effects of highly purified eicosapentaenoic acid on plasma beta-thromboglobulin level and vascular reactivity to angiotensin II. Artery 1987;14:295–303.
20 Knapp HR: Dietary omega-3 fatty acids and blood pressure control; in Drevon CA, Baksaas I, Krokan HE (eds): Omega-3 Fatty Acids: Metabolism and Biological Effects. Basel, Birkhäuser, 1993, pp 241–249.
21 Weisser B, Struck A, Gobel BO, et al: Fish oil and baroreceptor function in man. Klin Wochenschr 1990;68:49–52.
22 Carretta R, Fabris B, Bellini G, et al: Baroreflex function after therapy withdrawal in patients with essential hypertension. Clin Sci 1983;64:259–263.
23 Hintze TH, Martin EG, Messina EJ, et al: Prostacyclin (PGI$_2$) elicits reflex bradycardia in dogs: Evidence for vagal mediation. Proc Soc Exp Biol Med 1979;162:96–100.
24 Knapp HR: Hypotensive effects of omega-3 fatty acids: Mechanistic aspects. World Rev Nutr Diet. Basel, Karger, 1991, vol 66, pp 313–328.
25 Archer SL, Johnson GJ, Gebhard RL, et al: Effect of dietary fish oil on lung lipid profile and hypoxic pulmonary hypertension. J Appl Physiol 1989;66:1662–1673.
26 Knapp HR: Reduced allergen-induced nasal congestion and leukotriene synthesis with an orally active 5-lipoxygenase inhibitor. N Engl J Med 1990;323:1745–1748.
27 Morrow JD, Hill KE, Burk RF, et al: A series of prostaglandin F$_2$-like compounds are produced in vivo in humans by a non-cyclooxygenase, free radical-catalyzed mechanism. Proc Natl Acad Sci USA 1990;87:9383–9387.
28 Knapp HR, Miller AJ, Lawson JA: Urinary excretion of diols derived from eicosapentaenoic acids during n–3 fatty acid ingestion by man. Prostaglandins 1991;42:47–54.
29 Knapp HR: Altered excretion of eicosapentaenoic acid and arachidonic acid diols during the lowering of blood pressure by fish oil ingestion. Clin Res 1991;39:185A.

Howard R. Knapp, MD, PhD, Division of Clinical Pharmacology, Department of
Internal Medicine, University of Iowa, Iowa City, IA 52242 (USA)

Galli C, Simopoulos AP, Tremoli E (eds): Effects of Fatty Acids and Lipids in Health and Disease. World Rev Nutr Diet. Basel, Karger, 1994, vol 76, pp 15–20

....

Fatty Acids and Diabetes mellitus

Tonny Jensen

Steno Diabetes Center, Gentofte, Denmark

Careful dietary advice has always been important in the treatment of diabetic patients. The goal is to normalize not only blood glucose concentrations, but also other metabolic abnormalities including lipoprotein disorders commonly seen in such patients. It has been recommended that patients reduce saturated fat in their diets and replace it with complex carbohydrates [1]. However, recent studies have shown that this diet may not be the most effective approach to decreasing risk factors for coronary artery disease in all patients [2]. Other studies have demonstrated the need for different therapeutic approaches in type I and type II diabetic patients, and even within the two groups of patients advantages and disadvantages of dietary intervention may differ among individual patients.

In the following the effects of dietary intervention in type I (insulin-dependent) and type II (noninsulin-dependent) diabetic patients will be described separately. The purpose of this review is to discuss the perspectives of fatty acid supplementation in the prevention of atherosclerotic vascular disease in diabetic patients. The present knowledge of beneficial and potentially deleterious effects of monounsaturated and polyunsaturated fatty acids will be described and attempts to identify subgroups of patients in whom supplementation with these fatty acids may be most safely recommended, will be proposed.

Fatty Acid Supplementation in Type I Diabetic Patients

In 1986, Haines et al. [3] published the first detailed study on the effects of fish oil (MaxEpa) or olive oil supplementation in insulin-dependent diabetic patients. They found a significant reduction in platelet thromboxane production and a significant increase in low density lipoprotein (LDL) cholesterol during fish oil supplementation compared with olive oil supplementation. In a smaller study, Miller et al. [4] found no changes in LDL cholesterol, fasting blood glucose or glycosylated hemoglobin during dietary supplementation with 20 g MaxEpa daily for 8 weeks. Recently, two studies have demonstrated interesting effects of ω3 fatty acids on lipoprotein subfractions in male [5] and female [6] insulin-dependent diabetic patients. In the study by Mori et al. [5] high density lipoprotein-2 (HDL_2) cholesterol increased approximately 30% in 9 patients receiving 15 g MaxEpa daily. In that study, LDL cholesterol was unchanged and triglyceride was reduced by MaxEpa. Bagdade et al. [6] similarly found a substantial increase in HDL_2 cholesterol after intake of 6 g/day ω3 polyunsaturated fatty acids (as Super-Epa) in 8 normolipidemic women with type I diabetes. They found no changes in postheparin hepatic or lipoprotein lipase activities in fish-oil-treated subjects, suggesting that the observed changes in HDL_2 cholesterol were not consequences of increased lipase-mediated catabolism of very low density lipoprotein (VLDL).

Not all insulin-dependent diabetic patients are at the same risk for cardio-vascular disease. The increased mortality of cardiovascular disease is for the major part explained by an extremely high mortality in the subgroup of patients developing clinical nephropathy [7, 8]. These patients have a number of established cardiovascular risk factors such as atherogenic changes in plasma lipoproteins [9] and elevated blood pressure [10]. They also have signs of vascular vulnerability. The permeability of the whole vascular bed is clearly elevated in patients with incipient and clinical nephropathy [11]. Moreover, elevated plasma levels of von Willebrand factor and impaired fibrinolytic response to exercise have been demonstrated in these patients, suggesting that they have endothelial cell dysfunction or damage [12]. In a double-blind crossover study, we found that dietary supplementation with cod liver oil significantly reduced blood pressure from 146/90 to 139/85 mm Hg in such high-risk patients [13]. During cod liver oil supplementation the HDL cholesterol increased, VLDL cholesterol and triglyceride decreased and the elevated vascular permeability was partially normalized. No changes were observed in glomerular filtration rate, degree of albuminuria or glycemic control during supplementation with cod liver oil. In accordance with the findings by Spannagl et al. [14] we found no effects of cod liver oil on hemostatic parameters in insulin-dependent diabetic patients.

Thus, ω3 fatty acid supplementation does not seem to deterioriate glucose metabolism in type I diabetic patients. Moreover, the overall effects on lipoproteins and blood pressure of dietary marine oil supplementation seem beneficial especially in high-risk patients. Whether these effects, together with a possible protective effect on the vascular wall, will be of beneficial value in prevention of atherosclerosis in insulin-dependent diabetic patients can, however, only be settled in future long-term studies.

Fatty Acid Supplementation in Type II Diabetic Patients

The diet for noninsulin-dependent diabetic patients recommended by the American Diabetes Association (ADA) advises liberal intake of carbohydrates up to 60% of total energy and restriction of total fat to 30% of total energy. Saturated fatty acids are limited to <10% of total energy and cholesterol to <300 mg/day.

Most type II patients have an increase in VLDL triglyceride and a decrease in HDL cholesterol. These changes have been associated with an increased risk of atherosclerosis in this population [2, 15]. However, there is no evidence to believe that they will be changed favorably by the diet recommended. Indeed, such a diet has been shown to increase plasma VLDL triglyceride and lower HDL cholesterol in persons without diabetes as well as in persons with noninsulin-dependent diabetes mellitus. The study by Garg et al. [2] revealed that a diet high in monounsaturated fatty acids (from olive oil) improved glycemic control, reduced triglyceride and VLDL cholesterol levels and raised HDL cholesterol levels compared with a high carbohydrate diet. The authors therefore suggest that partial replacement of dietary carbohydrates with monounsaturated fatty acids may be beneficial in certain groups of type II patients (e.g. hypertriglyceridemic patients, those with HDL cholesterol levels <0.9 mmol/l, and elderly patients with poor compliance with high carbohydrate diets).

That type II diabetes mellitus is often associated with hypertriglyceridemia suggests a close relationship between hypertriglyceridemia and insulin resistance [16] and between glucose and lipid metabolism [17]. Since reductions in triglyceride and VLDL cholesterol have been observed in most fish oil trials, the metabolic effects of dietary ω3 fatty acid supplementation have also been studied in type II diabetic patients. A triglyceride-lowering effect of fish oil supplementation was found in three separate studies [18–20], whereas in one study ω3 fatty acids decreased serum triglyceride only in patients with hypertriglyceridemia [21]. However, in two studies the serum apoprotein B concentration significantly increased after fish oil supplementation [18, 21]. Because LDL apoprotein B levels correlate directly with the presence of coronary artery disease, these elevations may be unfavorable [22].

Other adverse metabolic effects of ω3 fatty acids have been found in noninsulin-dependent diabetes mellitus. Glauber et al. [19] found elevated hepatic glucose output and impaired insulin secretion but unchanged peripheral glucose disposal in 6 noninsulin-dependent diabetic men after intake of 18 g fish oil concentrate daily for 1 month. Similarly, Vessby et al. [23] found elevated fasting and postprandial blood glucose concentrations after a diet rich in ω3 fatty acids. Despite the elevated blood glucose concentrations, serum insulin levels were suppressed suggesting impaired insulin response to glucose after supplying a diet rich in ω3 fatty acids.

In a recent elegant study, McVeigh et al. [24] studied vascular response to acetylcholine infused into the brachial artery in noninsulin-dependent diabetic patients after dietary supplementation with fish oil or olive oil. Their principal finding was improvement, after intake of fish oil, of the endothelium-dependent response to acetylcholine by increasing stimulated nitric oxide release from the endothelium. Thus, also in noninsulin-dependent diabetic patients, fish oils seem to have a direct effect on the vascular wall of possible importance in protection against vasospasm and thrombosis.

At present it is too early to answer the question, whether a high intake of ω3 fatty acids is advantageous for type II diabetic patients. On the one hand there is a tendency to impairment of blood glucose homeostasis and insulin secretion, while on the other the effects on lipoproteins and the vascular wall may possibly outweigh this potentially negative effect with regard to the risk for accelerated atherosclerosis.

Future Perspectives and Recommendations of ω3 Fatty Acids in Diabetes mellitus

In the future, more studies on the effects of ω3 fatty acids on the development of microvascular complications are expected. Animal studies have suggested that fish oils, either alone or together with other polyunsaturated fatty acids, may have protective effects on the development of diabetic cardiomyopathy [25] and progression of diabetic renal disease [26]. In type I diabetic patients with nephropathy, long-term controlled studies of the effects of fish oil supplementation on prognosis and mortality are awaited. In contrast, it is still premature to start long-term intervention studies of dietary fish oil supplementation in type II diabetic patients. Before long-term studies can be initiated we need to know more about the biochemical and metabolic consequences of increased intake of fish oil in these patients.

References

1   American Diabetes Association: Nutritional recommendations and principles for individuals with diabetes mellitus. Diabetes Care 1986;10:126–132.
2   Garg A, Bonanome A, Grundy SM, et al: Comparison of a high-carbohydrate diet with a high-monounsaturated-fat diet in patients with non-insulin-dependent diabetes mellitus. N Engl J Med 1988;319:829–834.
3   Haines AP, Sanders TAB, Imesom JD, et al: Effects of a fish oil supplement on platelet function, haemostatic variables and albuminuria in insulin-dependent diabetics. Thromb Res 1986;43:643–655.
4   Miller ME, Anagnostou AA, Ley B, et al: Effects of fish oil concentrates on hemorheological and hemostatic aspects of diabetes mellitus: A preliminary study. Thromb Res 1987;47:201–214.
5   Mori TA, Vandongen R, Masarei JRL, et al: Comparison of diets supplemented with fish oil or olive oil on plasma lipoproteins in insulin-dependent diabetics. Metabolism 1991;40:241–246.
6   Bagdade JD, Buchanan WE, Levy RA, et al: Effects of omega-3 fish oils on plasma lipids, lipoprotein composition, and postheparin lipoprotein lipase in women with IDDM. Diabetes 1990;39:426–431.
7   Borch-Johnsen K, Kreiner S: Proteinuria: value as predictor of cardiovascular mortality in insulin-dependent diabetes mellitus. Br Med J 1987;294:1651–1654.
8   Jensen T, Borch-Johnsen K, Kofoed-Enevoldsen A, et al: Coronary heart disease in young type I (insulin-dependent) diabetic patients with and without diabetic nephropathy: Incidence and risk factors. Diabetologia 1987;30:144–148.
9   Jensen T, Stender S, Deckert T: Abnormalities in plasma concentrations of lipoproteins and fibrinogen in type I (insulin-dependent) diabetic patients with increased urinary albumin excretion. Diabetologia 1988;31:142–145.
10 Feldt-Rasmussen B, Borch-Johnsen K, Mathiesen ER: Hypertension in diabetes as related to nephropathy. Early blood pressure changes. Hypertension 1985;7(suppl II):18–20.
11 Feldt-Rasmussen B: Increased transcapillary escape rate of albumin in type I (insulin-dependent) diabetic patients with microalbuminuria. Diabetologia 1986;29:282–286.
12 Jensen T, Bjerre-Knudsen J, Feldt-Rasmussen B, et al: Features of endothelial dysfunction in early diabetic nephropathy. Lancet 1989;i:461–463.
13 Jensen T, Stender S, Goldstein K, et al: Partial normalization by dietary cod liver oil of increased microvascular albumin leakage in patients with insulin-dependent diabetes and albuminuria. N Engl J Med 1989;321:1572–1577.
14 Spannagl M, Drummer C, Froschl H, et al: Plasmatic factors of haemostasis remain essentially unchanged except for PAI activity during n-3 fatty acid intake in type I diabetes mellitus. Blood Coagul Fibrinolysis 1991;2:259–265.
15 Welborn TA, Knuiman M, McCann V, et al: Clinical macrovascular disease in Caucasoid diabetic subjects: Logistic regression analysis of risk variables. Diabetologia 1984;27:568–573.
16 Steiner G, Morita S, Vranic M: Resistance to insulin but not to glucagon in lean human hypertriglyceridemics. Diabetes 1980;29:899–905.
17 Lillioja S, Bogardus C, Mott DM, et al: Relationship between insulin-mediated glucose disposal and lipid metabolism in man. J Clin Invest 1985;75:1106–1115.
18 Schectman G, Kaul S, Kissebah AH: Effect of fish oil concentrate on lipoprotein composition in NIDDM. Diabetes 1988;37:1567–1573.
19 Glauber H, Wallace P, Griver K, et al: Adverse metabolic effect of omega-3 fatty acids in non-insulin-dependent diabetes mellitus. Ann Intern Med 1988;108:663–668.
20 Friday KE, Childs MT, Tsunehara CH, et al: Elevated plasma glucose and lowered triglyceride levels from omega-3 fatty acid supplementation in type II diabetes. Diabetes Care 1989;12:276–281.
21 Kasim SE, Stern B, Khilnani S, et al: Effects of omega-3 fish oils on lipid metabolism, glycaemic control, and blood pressure in type II diabetic patients. J Clin Endocrinol Metab 1988;67:1–5.
22 Avogaro P, Bon GB, Cazzolato G, et al: Are apolipoproteins better determinators than lipids for atherosclerosis? Lancet 1979;i:901–903.
23 Vessby B, Karlstrom B, Boberg M, et al: Polyunsaturated fatty acids may impair blood glucose control in type 2 diabetic patients. Diabetic Med 1992;9:126–133.

24 McVeigh GE, Brennan GM, Johnston GD, et al: Dietary fish oil augments nitric oxide production or release in patients with type 2 (non-insulin-dependent) diabetes mellitus. Diabetologia 1993; 36:33–38.
25 Black SC, Katz S, McNeill JH: Cardiac performance and plasma lipids of omega-3 fatty acid-treated streptozocin-induced diabetic rats. Diabetes 1989;38:969–974.
26 Barcelli UO, Weiss M, Beach D, et al: High linoleic acid diets ameliorate diabetic nephropathy in rats. Am J Kidney Dis 1990;16:244–251.

Tonny Jensen, MD, DMSci, Steno Diabetes Center, DK–2820 Gentofte (Denmark)

Galli C, Simopoulos AP, Tremoli E (eds): Effects of Fatty Acids and Lipids in
Health and Disease. World Rev Nutr Diet. Basel, Karger, 1994, vol 76, pp 21–22

....

Summary Statement: ω3 and ω6 Fatty Acids, Lipids and Lipoproteins

W.S. Harris

The session was co-chaired by *W.S. Harris* and *G. Crepaldi,* and presentations were made by Drs. *Harris, C. A. Drevon, K. C. Hayes, C. R. Sirtori, G. Crepaldi,* and *P. J. Nestel.*

Twenty years ago the effects of dietary fats on plasma lipid levels was quite simple: animal fats raised cholesterol levels, vegetable oils lowered them, monounsaturated oils had no effect, and fish oils (in the form of cod liver oil) were good for preventing rickets.

The presentations in this session made it quite clear that the situation is not so simple. The speakers discussed the impact of PUFA (especially ω3 fatty acids) on the metabolism of chylomicrons, VLDL, LDL, and HDL, the use of PUFA in the management of hyperlipidemia, and future research directions. The speakers emphasized several relationships between dietary PUFA and lipoprotein metabolism which appear to be relatively clear. A few of these were:

Saturated fatty acids containing 12, 14, and 16 carbon atoms are all hypercholesterolemic, but to different extents depending on the experimental conditions. Saturated fatty acids containing less than 12 or more than 16 carbon atoms are not hypercholesterolemic.

The consumption of linoleic acid within the normal range for usual diets (5–8% energy) does not lower HDL cholesterol levels.

Practical intakes of ω3 fatty acids (2–4 g/day) lower fasting and postprandial triglyceride levels, raise HDL2 levels but do not significantly alter LDL cholesterol levels. The resultant impact on cardiovascular risk is unclear.

There, of course, remain many areas of continuing controversy:

Are monounsaturated or polyunsaturated fatty acids the best substitutes for saturated fatty acids in the diet – both in terms of effects on lipoprotein *levels* and on *oxidation susceptibility?*

Does a relatively high fat diet rich in oleic acid (the Mediterranean diet) lower cardiovascular risk more than a diet low in total and saturated fats?

What are the biochemical mechanisms by which fatty acids alter serum lipoprotein levels?

Interest in the area of dietary lipids, serum lipoproteins and coronary disease remains very high. The potential beneficial roles played by ω6 and ω3 PUFA will continue to stimulate research in the years to come.

Galli C, Simopoulos AP, Tremoli E (eds): Effects of Fatty Acids and Lipids in
Health and Disease. World Rev Nutr Diet. Basel, Karger, 1994, vol 76, pp 23–25

....

Chylomicron Metabolism and ω3 and ω6 Fatty Acids

William S. Harris

Lipid and Arteriosclerosis Prevention Clinic, Division of Clinical Pharmacology,
Department of Medicine, University of Kansas Medical Center,
Kansas City, Kans., USA

Chylomicrons are triglyceride-rich lipoproteins made in the intestine, secreted into the lymphatics and carried into the blood. They carry dietary fats to body tissues which either burn them for energy or store them for future use. Once the chylomicrons deliver the fat they contain, they are called 'chylomicron remnants'. These relatively cholesterol-enriched particles are cleared by the liver. High levels of chylomicron remnants are believed to be atherogenic [1]. Different polyunsaturated fatty acids (PUFAs) alter the way the body produces and/or metabolizes chylomicrons, with the ω3 fatty acids having a greater impact than the ω6.

Absorption of ω3 and ω6 PUFAs

Most fatty acids are absorbed quite well, the only exceptions being the longer chain saturates (stearic, arachidic) which can form calcium soaps in the gut, reducing their absorption. Fatty acids of the ω6 class are readily absorbed, but those of the ω3 class appear to be somewhat resistant to digestion/absorption [2]. Nevertheless, when taken in small quantities along with other foods, all of the fatty acids are ultimately absorbed.

Supplemental ω3 and ω6 PUFA and Fat Absorption

When ω3 fatty acids are present in the background diet, some aspect of chylomicron metabolism is altered resulting in lower postprandial lipid levels in response to a normal fat load (i.e., a meal containing usual fats – not fish oil). Harris and Connor [3] were the first to observe this effect, but it was unclear from that study whether the effect was due to fish oil in the test meal or fish oil in the background diet. Subsequent studies by Harris et al. [4] and by Weintraub et al. [5] confirmed that ω3 fatty acids in the background diet were responsible for the reduced chylomicron levels; there was no reduction when subjects on a normal diet consumed a fish oil test meal.

This effect first observed with daily intakes of 26 g [4] or 8 g [5] of ω3 fatty acids (79 or 27 g of fish oil concentrate/day). These are unrealistic intakes. At least three studies [6–8] have now been done using smaller, more achievable intakes of ω3 fatty acids (1–4 g/day), and each has found the same effect: significantly reduced (20–40%) postprandial lipemia.

Possible Mechanisms

Although the effect has been observed by several groups, the mechanisms responsible are not known. There are only two possible ways to lower plasma levels of chylomicrons: reducing their rate of entry into the blood or increasing their rate of removal from it. There is little solid evidence in humans for either.

Lipoprotein lipase and hepatic triglyceride lipase are the two intravascular enzymes responsible for clearing chylomicron triglyceride from the bloodstream. Although extensively studied, no increased activity of either of these enzymes has been found when assayed in post-heparin plasma [4, 5, 9]. In addition, chylomicrons obtained after a normal fat load from subjects consuming fish oil in their background diets are not metabolized more rapidly by lipoprotein lipase in vitro than are control chylomicrons [5]. Finally, when chylomicron-like particles are injected intravenously into subjects taking ω3 fatty acid supplements, they are not removed more rapidly than usual [8]. These three observations argue rather strongly against fish oils accelerating chylomicron clearance rates.

If chylomicrons are not removed more rapidly, then the only option remaining is slower secretion into the blood. Cultured intestinal cells were reported to synthesize and secrete less triglyceride after incubation with ω3 fatty acids than with oleic acid [10]. However, slowed fat absorption was not observed in lymph-cannulated rats chronically fed fish oil [11]. Therefore, the mechanism of the ω3 fatty acid-induced reduction in postprandial lipemia remains obscure. But regardless of the mechanism, lower chylomicron and remnant levels may be considered to be potentially antiatherogenic.

References

1   Zilversmit DB: Atherogenesis: A postprandial phenomenon. Circulation 1979;3:473–485.
2   Yang L-Y. Kuksis A, Myher JJ: Lumenal hydrolysis of menhaden and rapeseed oils and their fatty acid methyl and ethyl esters in the rat. Biochem Cell Biol 1989;67:192–204.
3   Harris WS, Connor WE: The effects of salmon oil upon plasma lipids, lipoproteins and triglyceride clearance. Trans Assoc Am Physicians 1980;43:148–155.
4   Harris WS, Connor WE, Alam N, Illingworth DR: Reduction of postprandial triglyceridemia in humans by dietary ω3 fatty acids. J Lipid Res 1988;29:1451–1460.
5   Weintraub MS, Zechner R, Brown A, et al: Dietary polyunsaturated fats of the ω6 and ω3 series reduce postprandial lipoprotein levels. J Clin Invest 1988;82:1884–1893.
6   Brown AJ, Roberts DCK: Moderate fish oil intake improves lipemic response to a standard fat meal. Arterioscler Thromb 1991;11:457–466.
7   Harris WS, Windsor SL: n-3 fatty acid supplements reduce chylomicron levels in healthy volunteers. J Appl Nutr 1991;43:5–15.
8   Harris WS, Muzio F: Fish oil reduces postprandial triglyceride concentrations without accelerating lipid emulsion removal rates. Am J Clin Nutr 1993;58:68–74.
9   Nozaki S, Garg A, Vega GL, Grundy SM: Post-heparin lipolytic activity and plasma lipoprotein response to omega-3 polyunsaturated fatty acids in patients with primary hypertriglyceridemia. Am J Clin Nutr 1991;53:638–642.
10 Ranheim T, Gedde-Dahl A, Rustan AC, et al: Influence of eicosapentaenoic acid (20:5, n-3) on secretion of lipoproteins in CaCo-2 cells. J Lipid Res 1992;33:1281–1293.
11 Herzberg GR, Chernenko GA, Barrowman JA, et al: Intestinal absorption of fish oil in rats previously adapted to diets containing fish oil or corn oil. Biochim Biophys Acta 1992;1124:190–194.

William S, Harris, PhD, Lipid and Arteriosclerosis Prevention Clinic,
Division of Clinical Pharmacology, Department of Medicine,
University of Kansas Medical Center, Kansas City, KS 66160-7418 (USA)

Galli C, Simopoulos AP, Tremoli E (eds): Effects of Fatty Acids and Lipids in
Health and Disease. World Rev Nutr Diet. Basel, Karger, 1994, vol 76, pp 26–29

....

VLDL Metabolism and ω3 Fatty Acids

C.A. Drevon[a], *A.C. Rustan*[b]

[a] Institute of Nutrition Research and [b] Institute of Pharmacy,
University of Oslo, Norway

From studies in primary cultures of rat hepatocytes we have shown that secretion of VLDL is reduced in the presence of several long-chain polyunsaturated fatty acids, whereas oleic acid and palmitic acid are the most potent stimulators of synthesis and secretion of triglycerides (TG) to the medium [1]. Both eicosapentaenoic acid (EPA) and docosahexaenoic acid (DHA) promote a low synthesis of TG, with EPA being the least stimulatory fatty acid of all tested. In cultured hepatocytes the synthesis of TG is low in the presence of EPA, whereas the secretion of VLDL per se is unaffected [2]. In our acute experiments we could not find any increase in fatty acid oxidation [2]. In whole animal experiments we observed that dietary supplementation with EPA and DHA caused a markedly lower postprandial plasma concentration of TG, free glycerol and free fatty acids (FFA), as compared to animals fed saturated fatty acids [3]. In nonfasting conditions the formation of TG takes place in the intestine as well as the liver. There is some evidence from human in vivo studies that the secretion of chylomicrons is decreased with very long-chain ω3 fatty acids [4]. We examined the effect of feeding rats ω3 fatty acids, as highly concentrated EPA/DHA, on fatty acid metabolism in adipose tissue, muscle, liver and on the whole body.

Materials and Methods

The ω3 fatty acid concentrate (K85) was donated by Pronova AS, Oslo, Norway. Lard was obtained from Agro Fellesslakteri, Oslo, Norway. Male Wistar rats weighing 165–180 g were pair-fed ω3 fatty acids or lard-supplemented diets for periods up to 8 weeks. They were randomly divided into two groups and transferred to individual cages [5]. Each animal group was offered one of two semisynthetic diets, L, lard (19.5% lard) or F, fish oil derived ω3 fatty acids (13% lard and 6.5% K85), and an additional 1.5% soybean oil to both dietary groups. The rats were fed 20 g/day of fresh experimental diets, and the actual food intake was similar in both groups. Body weight was registered twice weekly. The rats had access to food only during darkness, to ensure that the animals were in a fed state at 09.00 h the next day. Two to 6 rats from each group were decapitated and plasma and fat pads isolated, after 2 days, and 1, 2, 3, 4 and 6–8 weeks of the feeding regimen, respectively. In some experiments, 9 rats in each group were used for studies of whole body metabolism by indirect calorimetry. By measurement of O_2 consumption and CO_2 production, the mean respiratory quotients (RQ) were determined. The substrate utilization (fat and carbohydrate oxidation) during the experiment (22 h) was calculated from the calorimetric data. Blood was collected by decapitation with 0.1% EDTA and immediately chilled on ice. Plasma was prepared and stored at –70 °C prior to analysis of triacylglycerol, FFA and glycerol [1, 2].

Adipose tissue samples (1.5–2 g) were taken from epididymal, perirenal, mesenteric (abdominal) and subcutaneous fat pads, and adipocytes were isolated [6, 7]. Assuming that the bulk of the cells from adipose tissue is made up of triacylglycerol, and that the cells are spherical in shape, the volume was calculated from the radius. The size of the adipocytes was measured in aliquots of isolated cells using a microscope fitted with a graduated ocular. The cells were observed under low magnification with the aid of a calibrated grid and the mean diameter of 120–150 cells from each cell preparation was calculated.

Lipolysis from adipose cells was measured as the amount of glycerol released by freshly isolated fat cells after 2 h incubation at 37 °C. Adipocytes corresponding to 10–60 mg lipid were incubated in a total volume of 2 ml buffer containing 5 mM glucose and 3% f-BSA, and either no isoprenaline (basal lipolysis) or 0.1 μM isoprenaline (total lipolysis). The incubations were terminated by placing the samples on ice. The glycerol content in the medium was determined fluorimetrically [5]. A blank for each fat tissue was placed on ice immediately after preparation without incubation. Lipid content of adipocytes was determined gravimetrically after extraction three times with 5 volumes of hexane.

Results and Discussion

After 7–8 weeks of feeding, plasma concentration of TG, phospholipids, free glycerol and FFA was decreased in rats fed ω3 fatty acids as compared to rats fed lard. Plasma concentration of glucose was somewhat higher in animals fed highly concentrated EPA/DHA, whereas there was no significant difference in plasma ketone body concentration. The concentration of TG in the liver was similar for animals in both dietary groups. By indirect calorimetry we could show that fasted and nonfasted rats fed ω3 fatty acids for 3–5 weeks had an increased oxidation of carbohydrates, and reduced oxidation of fatty acids. Total energy expenditure, body weight and fat absorption were similar in both feeding groups. Plasma concentration of glucose was slightly increased in animals fed ω3 fatty acids, whereas there was no significant difference in plasma concentration of ketone bodies and insulin. Glycogen concentration in liver and skeletal muscle was significantly lower in ω3 fatty acid-fed animals as compared to lard-fed rats. There was a significantly lower secretion of labeled TG from freshly isolated hepatocytes from animals fed ω3 fatty acids, whereas there was no difference in the amount of TG in hepatocytes isolated from rats in both dietary groups [5].

The ω3 fatty acid diet increased peroxisomal fatty acid oxidation (measured as hydrogen peroxide production) in isolated hepatocytes by approximately 100% using oleic acid or erucic acid as substrate [5]. Mitochondrial fatty acid oxidation was similar in both dietary groups as evaluated by ketone body formation in isolated hepatocytes from animals fed the different diets for 6–8 weeks. Total fatty acid oxidation (peroxisomal and mitochondrial) was similar in the two feeding groups as the peroxisomal fatty acid oxidation only accounts for 10–20% of total fatty acid oxidation.

The basal lipolysis in epididymal adipocytes was decreased in animals fed ω3 fatty acids as compared to animals fed lard. We also observed that adipocytes isolated from rats fed ω3 fatty acids were somewhat more sensitive to isoproterenol, causing increased lipolysis. Similarly data have been reported by Parrish et al. [8]. Furthermore, perirenal and epididymal adipocytes from rats fed EPA/DHA were significantly reduced in size, compared to the size of adipocytes from animals fed lard, whereas there was no difference in the size of adipocytes from mesenteric and subcutaneous sources. We also found that incorporation of radiolabeled glucose into TG in isolated epididymal cells incubated with oleic acid or EPA, was similar irrespective of the diet fed to the rats. These findings suggest that ω3 fatty acids do not execute their primary effect on plasma TG via the adipose tissue.

Conclusion

In whole animal experiments there is a marked lowering of nonfasting plasma concentration of TG, phospholipids, FFA and free glycerol after feeding ω3 fatty acid-enriched diet, as compared to lard feeding. The secretion of TG from hepatocytes isolated from animals fed ω3 fatty acids was significantly reduced, whereas hepatic TG concentration was similar in both feeding groups. We also observed that whole body fatty acid oxidation was reduced in rats fed ω3 fatty acids, and that the size of adipocytes in epididymal and perirenal fat tissue was markedly decreased on this diet. Our findings were somewhat surprising based on the fact that some papers report increased hepatic β-oxidation [3]. However, it should be noted that our study [5] is the first to examine whole body energy metabolism in animals fed very long-chain ω3 fatty acids.

At present we are exploring if the reduced plasma concentration of FFA provides less substrate for hepatic and adipose synthesis of TG and thereby promotes low concentration of plasma TG, low accumulation of TG in some adipose tissues and reduced whole body fatty acid oxidation. Based on these observations, we suggest that the reduced level of plasma unesterified fatty acids and glycerol after feeding ω3 fatty acids in the postprandial state to a large extent is caused by reduced production/or secretion of triacylglycerol-rich lipoprotein particles by the intestine [5, 9].

References

1   Nossen JØ, Rustan AC, Gloppestad SH, Målbakken S, Drevon CA: Eicosapentaenoic acid inhibits synthesis and secretion of triacylglycerols by cultured rat hepatocytes. Biochim Biophys Acta 1986;879:56–65.
2   Rustan AC, Nossen JØ, Christiansen EC, Drevon CA: Eicosapentaenoic acid reduces hepatic synthesis and secretion of triacylglycerol by decreasing the activity of acyl-coenzyme A:1,2-diacylglycerol acyltransferase. J Lipid Res 1988;29:1417–1426.
3   Rustan AC, Christiansen EN, Drevon CA: Serum lipids, hepatic glycerolipid metabolism and peroxisomal fatty acid oxidation in rats fed n-3 and n-6 fatty acids. Biochem J 1992;283:333–339.
4   Harris WS: Fish oils and plasma lipid and lipoprotein metabolism in humans: A critical review. J Lipid Res 1989;30:785–807.
5   Rustan AC, Hustvedt B-E, Drevon CA: Dietary supplementation of very long-chain n-3 fatty acids decreases whole body lipid utilization and reduces lipolysis from adipose tissue in the rat. J Lipid Res 1993;34:1299–1309.
6   Rodbell M: Metabolism of isolated fat cells. I. Effects of hormones on glucose metabolism and lipolysis. J Biol Chem 1964;239:375–380.
7   Di Girolamo M, Mendinger MS, Fertig J: A simple method to determine fat cell size and number in four mammalian tissues. Am J Physiol 1971;221:850–858.
8   Parrish CC, Pathy DA, Parkes JG, Angel A: Dietary fish oils modify adipocyte structure and function. J Cell Physiol 1991;148:493–502.
9   Harris WS, Muzio F: Fish oil reduces postprandial triglyceride levels without accelerating lipid-emulsion removal rates. Am J Clin Nutr 1993;58:68–74.

Prof. Christian A. Drevon, MD, PhD, Section for Dietary Research,
Institute for Nutrition Research, University of Oslo, PO Box 1117 Blindern,
N–0317 Oslo (Norway)

Galli C, Simopoulos AP, Tremoli E (eds): Effects of Fatty Acids and Lipids in
Health and Disease. World Rev Nutr Diet. Basel, Karger, 1994, vol 76, pp 30–33

....

Low Density Lipoprotein Metabolism and ω3 Fatty Acids

K.C. Hayes

Foster Biomedical Research Laboratory, Brandeis University, Waltham, Mass., USA

The ability of certain dietary fatty acids to affect plasma lipid levels is well
known [1]. In humans, most of this effect on lipids is mediated through changes
in the circulating low density lipoprotein cholesterol (LDL-C) pool, with high
density lipoprotein cholesterol (HDL-C) modestly increasing during saturated
fat intake or decreasing during periods of high ω6 polyunsaturated fatty acid
(PUFA) intake (>20%en as 18:2).

An overall scheme for how this might occur is depicted in figure 1. This
model indicates that the lipoprotein pool depends in large part on the initial
synthesis of hepatic triglycerides and secretion of very low density lipoprotein
(VLDL), which is affected both by dietary fat composition and the amount of
cholesterol consumed. Catabolism of VLDL is important because its surface
coat is a major source of HDL whereas most of the remnant clears via the LDL_r,
providing the latter are not down-regulated. Decreased LDL_r activity leads to
diversion of VLDL remnants to LDL, causing expansion of that pool. The LDL
pool would expand further in this situation if VLDL production is elevated or if
cholesteryl ester transfer protein (CETP) activity is accelerated, both of which
can result from dietary cholesterol loading, which also tends to depress LDL_r
activity. On the other hand, the LDL_r can be enhanced if bile acid synthesis is
increased to facilitate removal of hepatic cholesterol.

The impact of highly unsaturated fatty acids (HUFA) i.e. ω3 HUFAs such
as eicosapentaenoic acid (EPA, 20:5ω3) and docosahexaenoic acid (DHA,
22:6ω3) found in greatest concentrations in fish oil, have been examined
extensively for their effect on lipid metabolism over the past 10 years [2]. It is
generally agreed that EPA and DHA are effective agents for lowering plasma
triglycerides when supplemented at a minimal level of 1–3 g/day, or about

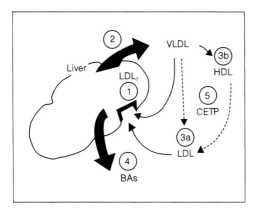

Fig. 1. Metabolic key to dietary impact on lipoprotein metabolism. 1 = LDL$_r$ activity is key; 2 = VLDL output; 3a = LDL formation; 3b = HDL formation; 4 = bile acid synthesis; 5 = CETP adds to LDL [from 1].

0.6%en, either incorporated in fish oil or as concentrated isolates or ethyl esters. However, it is also apparent that the total plasma cholesterol and LDL-C as well as apolipoprotein B (apoB) levels do not necessarily decline [3]. They may even increase. The increase is not thought to reflect increased mass of LDL conversion from VLDL because ω3 HUFAs depress VLDL synthesis and secretion. Kinetic studies with ^{125}I-LDL further suggest that LDL production is depressed [2]. Some discrepancies exist from animal studies where much higher intakes of fish oil (up to 20–40% of the dietary energy as fish oil) are fed. However, data collected under such conditions is of questionable relevance to human clinical studies because normal fatty acid metabolism depends on relationships between simultaneously ingested fatty acids [1]. Thus, extreme supplements of any given fatty acid can lead to spurious results if specific key fatty acids (especially ω6 PUFAs) are displaced by the supplementation. Also, phospholipid composition and metabolism vary greatly among species, so extrapolation of fatty acid effects from lower mammals to humans must be made with caution.

We recently had occasion to examine these relationships in humans given methyl esters of EPA (3.6 g) and DHA (2.9 g) as a daily concentrate [3]. Prior to receiving the supplement, LDLs were isolated from 7 normolipemic individuals and incubated with HEP-G2 cells to determine their impact on LDL receptor activity as well as key regulatory mRNA levels for proteins involved in cholesterol metabolism. After 2 weeks of supplementation, the LDLs were again isolated and characterized for their interaction with HEP-G2 cells.

A number of responses were noteworthy. Plasma lipids revealed a decrease in triglycerides of 30% in 2 weeks, even though plasma cholesterol and LDL-C

Table 1. Baseline cholesterol and triglyceride values and LDL fatty acid profiles are compared with values after 2 weeks of 7 g/day ω3 HUFAs

	Baseline	Fish oil
Total cholesterol, mg/dl	174 ± 24	172 ± 24
LDL-C	103 ± 20	106 ± 23
HDL-C	49 ± 8	51 ± 7
Total triglycerides, mg/dl	105 ± 81	74 ± 29[a]
LDL fatty acids, %		
Total ω6	49 ± 2	41 ± 4[a]
Total ω3	2.9 ± 0.5	10.3 ± 2[a]

[a] Value (mean ± SD) significantly lower than baseline (p<0.05).

did not change (table 1). The LDL particle became slightly less dense as the cholesterol ester/protein ratio increased 34%. This was associated with a 5-fold increase in 20:5ω3 and doubling of 22:6ω3 fatty acids in the fish oil-LDL particles.

Incubation of HEP-G2 cells with fish oil-LDL caused a striking and rapid decrease in LDL receptor activity attributed to a reduction in receptor number to about 1/3 normal. The particles demonstrated high binding affinity for receptors and rapid internalization evidenced by the fact that LDL$_r$ mRNA and apoA1 mRNA abundances were rapidly depressed, even at very low concentrations of LDL. The decrease in LDL density was highly correlated both with the rise in the ω3/ω6 fatty acid ratio among LDL lipids and with the decrease in LDL$_r$ number.

From the various sources of available information, it would appear that EPA and DHA have the ability to increase hepatic fatty acid oxidation, decrease fatty acid and triglyceride synthesis, and decrease VLDL secretion which leads to a drop in LDL production in direct proportion to the amount of ω3 HUFAs ingested. At the same time the LDL particle (and other lipoproteins as well) is modified compositionally, especially in its ω3 HUFAs, to the point that it becomes capable of down-regulating LDL receptor activity independent of any other lipoprotein or serum factor. The balance between depressed VLDL-LDL production and decreased LDL clearance is apparently rather tight such that ordinarily little change is found in total plasma LDL, even though total triglycerides are depressed. Because VLDL represents a major precursor of HDL, the decline in VLDL output is coupled with a depression in apoA1 synthesis and secretion by the liver leading eventually to depression in HDL if the ω3 HUFA intake is large and long enough.

Neither lipoprotein lipase nor hepatic triglyceride lipase are thought to be altered by ω3 HUFAs, so altered clearance of the lipoprotein triglyceride pool does not seem to be affected [2]. On the other hand, CETP activity is impaired by these long-chain HUFAs which would account for increases in HDL size as well as the often encountered relative *decreases* in cholesterol and phospholipid of the LDL particle and *increased* LDL density following long-term consumption of fish oil [2, 4]. The observation that long-term (4 months fish oil intake) was no different than long-term corn oil in terms of LDL impact on LDL_r activity in HEP-G2 cells and fibroblasts differs from our short-term crossover design as well as our findings [4]. Unfortunately the plasma lipid profiles of the 4 donor subjects for LDL binding studies for each oil were not provided. Thus, the relative 'expected activities' for LDL particles could not be related to the host subject's lipid metabolism, i.e. plasma lipids from the 4 subjects selected from each treatment group may have been comparable so that expectations for their LDL response would not differ. Nonetheless, short-term and long-term effects as well as individual variation may be somewhat dissimilar in terms of the LDL response. Further research will be needed to resolve these points.

References

1 Hayes KC, Khosla P: Dietary fatty acids and cholesterolemia. FASEB J 1992;6:2600–2607.
2 Nestel PJ: Effects of n3 fatty acids on lipid metabolism. Annu Rev Nutr 1990;10:149–168.
3 Lindsey S, Pronczuk A, Hayes KC: Low density lipoprotein from humans supplemented with n-3 fatty acids depresses both LDL receptor activity and LDL_r mRNA abundance in HepG2 cells. J Lipid Res 1992;33:647–658.
4 Nenseter MS, Rustan AC, Lund-Katz S, et al: Effect of dietary supplementation with n-3 polyunsaturated fatty acids on physical properties and metabolism of low density lipoprotein in humans. Arterioscler Thromb 1992;12:369–379.

K.C. Hayes, DVM, PhD, Foster Biomedical Research Laboratory, Brandeis University, Waltham, MA 02254 (USA)

Galli C, Simopoulos AP, Tremoli E (eds): Effects of Fatty Acids and Lipids in
Health and Disease. World Rev Nutr Diet. Basel, Karger, 1994, vol 76, pp 34–37

....

Effects of Unsaturated Fatty Acids on High Density Lipoprotein Metabolism

Cesare R. Sirtori

Center E. Grossi Paoletti and Institute of Pharmacological Sciences,
University of Milan, Italy

The interest in high density lipoprotein (HDL) levels and metabolism is well explained by the clear negative correlation between HDL, and particularly HDL-cholesterol (HDL-C) levels and the risk of coronary heart disease (CHD). While the raising effect of different drugs, e.g. fibric acids, is well established, concerns have been frequently raised about the potential of some dietary factors, in particular polyunsaturated (ω6) fatty acids (PUFA) to lower HDL-C levels [1].

Animal and Clinical Observations Related to Mechanisms

The majority of data pertaining to the intake of ω6 PUFA in the diet clearly indicates that both HDL and apolipoprotein AI (apoAI) levels tend to be decreased after these regimens [1]. Exceptions are studies with limited changes in the polunsaturate/saturate (P/S) ratio and/or with diets a low percent of fat (<30%) [2]. The mechanism of the HDL/apoAI reduction is apparently consequent to an inhibited apoAI synthesis by the liver, as well established by the determination of mRNA levels at this site, and by the observation of a reduced AI release following perfusion with different fatty acids [3]. The clear difference in the AI liver expression is confirmed by the comparative evaluation of PUFA versus monounsaturated fatty acid (MUFA, ω9) intake in the rat, these latter clearly inducing an enhanced AI mRNA expression [4]. Therefore, it can be safely concluded that the HDL-C reduction is a likely consequence of a

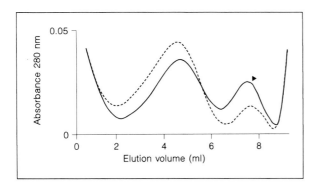

Fig. 1. Changes in the HDL subfraction distribution after ω3 EE intake in normal volunteers. A clear rise in the zonal ultracentrifugal fraction corresponding to HDL$_2$ (arrow) is noted after ω3 intake (–) [from 7].

reduced AI expression, related to translational mechanisms, possibly involving the peroxisome proliferator-associated receptor (PPAR) system, as also discussed below.

Clearly contrasting findings are seen with MUFA, generally represented by olive oil. In this case, in fact, the majority of studies either do not indicate any clear effect on the HDL system, or else show an increase of HDL-C levels [1]. In our own experience, the most characteristic consequence of olive oil intake is a rise in the apoAI/B ratio [5], also shown with other MUFA diets, e.g. erucic acid poor rapeseed oil, also eliciting a clear-cut increase of the apoAI/B ratio versus a saturated diet [6]. A MUFA diet enhances AI mRNA expression in rat hepatocytes versus a linoleic acid (LA)-rich diet. The authors attribute this finding to either increased transcription or mRNA stabilization [4]. Both in the case of oleic- and LA-rich diets, in our long-standing experience, definite changes in the HDL subfraction distribution in plasma versus saturated fat diets could not be established.

A clear difference in the *HDL subfraction distribution* is instead noted after ω3 fatty acid-rich diets. Experience with these diets varies in terms of the selected preparation for treatment (either fish oil, or fish oil concentrate or, more recently, ω3 ethyl esters (EE)). Our experience is that of the use of ω3 EE, generally given as 6 g/day of the classical K-85 preparation or of similar formulations. In man, ω3 EE do not generally change HDL levels but, in all cases, be they normolipidemic individuals or hypercholesterolemic patients, there is marked increase in the HDL$_2$/HDL$_3$ ratio in plasma [7]. This is well exemplified in figure 1, related to normolipidemics. The change in the HDL subfraction ratio occurs in the face of essentially unchanged HDL-C levels.

Table 1. ω3 fatty acids and HDL changes – differences between humans and primates

Humans:	No change of total HDL-C, increase of HDL$_2$
Primates:	Fall of total HDL, no apparent change in distribution
Reasons:	Fall of CETP and LCAT with both Lack of HL and lesser importance of CETP in primates make the effect on LCAT predominant

In order to explain these findings, a parallel evaluation of data coming from *primate experiments* is opportune. In primates, in fact, the plasma lipid reduction (mainly total cholesterol) occurring after fish oil intake (no data on ω3 EE is available) occurs to a large extent in the HDL fraction [8]. This is of particular interest since this potentially atherogenic change occurs while the animals clearly display a reduced atherosclerosis. When examining the lipoprotein metabolic changes after ω3 intake, two alterations are particularly evident. One is a reduction in the cholesteryl ester transfer protein (CETP) activity (not clearly of total levels), the other a reduction of lecithin cholesterol acyl transferase (LCAT) [9]. While the reduced CETP activity is described only for man, LCAT reduction takes place in both species [10].

These very interesting parallel observations in humans and primates point out to hypothetical mechanisms, whereby the intake of ω3 fatty acids may differently affect the lipoprotein pattern, and also possibly why the HDL subfraction distribution may be so dramatically altered in man. The enhanced peripheral triglyceride/VLDL catabolism exerted by ω3, possibly by way of a stimulated PPAR activity [11], will result, in the presence of inhibited CETP activity, in a rise of HDL$_2$ levels. HDL particles will in fact tend to become cholesteryl ester enriched, thus being larger, but with a reduced exchange for triglycerides from VLDL. In this way, HDL$_2$ in man will also be a poor substrate for hepatic lipase, i.e. resulting in a condition not different from that of nicotinic acid treatment [12]. In the case of primates, instead, the low levels of hepatic lipase activity will tend to make CETP nonfunctional and the marked reduction of LCAT will directly lead to a decreased formation of HDL particles (table 1).

Conclusions

The effect of unsaturated oils on the HDL system is surprisingly different for the three types of major fatty acids (ω3, ω6, ω9). While ω6 PUFAs tend to reduce HDL levels, mainly because of a negative impact on the apoAI liver expression, an ω9 (MUFA) diet will tend to raise HDL-C and apoAI for the opposite reason. Neither of the two seems to affect to a large extent the HDL subfraction distribution (possibly a LA-rich diet may somewhat reduce HDL$_2$). In contrast to these two more classical diets, the intake of ω3 EE will instead lead to a clear-cut rise of the HDL$_2$ subfraction, with no changes of total HDL-C concentrations. This effect is species-specific for man, since, e.g. in primates, total HDL levels are reduced. In the case of ω3, a combined inhibitory activity on CETP and LCAT, while in the presence of an enhanced peripheral catabolism of fatty acids, may explain both the changes in subfraction distribution in man, as well as the HDL reduction in primates.

References

1   Grundy SM, Denke MA: Dietary influences of serum lipids and lipoproteins. J Lipid Res 1990; 31:1149–1172.
2   Berry EM, Eisenberg S, Haratz D, et al: Effects of diets rich in monounsaturated fatty acids on plasma lipoproteins. The Jerusalem Nutrition Study: High MUFAs vs. PUFAs. Am J Clin Nutr 1991;53:899–907.
3   Sorci-Thomas M, Prack MM, Dashti N, et al: Differential effects of dietary fat on the tissue-specific expression of the apolipoprotein AI gene: Relationship to plasma concentrations of high density lipoproteins. J Lipid Res 1989;30:1397–1403.
4   Osada J, Fernandez-Sanchez A, Diaz-Morillo JL: Hepatic expression of apolipoprotein AI genes in rats is up-regulated by monounsaturated fatty acid diet. Biochim Biophys Res Commun 1991;180: 162–168.
5   Sirtori CR, Tremoli E, Gatti E, et al: Controlled evaluation of fat intake in the Mediterranean diet. Comparative activities of olive oil and corn oil on plasma lipids and platelets in high risk patients. Am J Clin Nutr 1986;44:635–642.
6   Valsta LM, Jauhiainen M, Aro A, et al: Effects of a monounsaturated rapeseed oil and a polyunsaturated sunflower oil diet on lipoprotein levels in humans. Atheroscler Thromb 1992;12: 50–57.
7   Franceschini G, Calabresi L, Maderna P, et al: ω-3 fatty acids selectively raise high-density lipoprotein 2 levels in healthy volunteers. Metabolism 1991;40:1283–1286.
8   Parks JS, Kaduck-Sawyer J, Bullock BC, et al: Effect of dietary fish oil on coronary artery and aortic atherosclerosis in African green monkeys. Arteriosclerosis 1990;10:1102–1112.
9   Abbey M, Clifton P, Kestin M, et al: Effect of fish oil on lipoproteins, lecithin-cholesterol acyltransferase, and lipid transfer protein activity in humans. Arteriosclerosis 1990;10:85–94.
10 Parks JS, Rudel LL: Effects of fish oil on atherosclerosis. Atherosclerosis 1990;84:83–94.
11 Rustan AC, Christiansen EN, Drevon CA: Serum lipids, hepatic glycerolipid metabolism and peroxisomal fatty acid oxidation in rats fed ω-3 and ω-6 fatty acids. Biochem J 1992;283:333–339.
12 Taskinen M-R, Nikkilä EA: Effects of acipimox on serum lipids, lipoproteins and lipolytic enzymes in hypertriglyceridemia. Atherosclerosis 1988;69:249–255.

Cesare R. Sirtori, MD, PhD, Center E. Grossi Paoletti,
Via Balzaretti, 9, I-20133 Milano (Italy)

Galli C, Simopoulos AP, Tremoli E (eds): Effects of Fatty Acids and Lipids in
Health and Disease. World Rev Nutr Diet. Basel, Karger, 1994, vol 76, pp 38–40

....

Unsaturated Fatty Acids in the Management of Hyperlipidemias

Gaetano Crepaldi, Sabina Zambon, Enzo Manzato

Department of Internal Medicine, University of Padova, Italy

The dietary content in saturated fatty acids has for a long time been associated with the plasma cholesterol levels and the incidence of coronary heart disease [1]. Several prevention trials, both primary and secondary, assessed the value of dietary modifications in preventing coronary heart disease [2]. In earlier studies the dietary modification was usually a reduction in saturated fatty acids, with a considerable increase in the polyunsaturated to saturated (P/S) ratio. In several secondary prevention trials the dietary modifications were included in other types of intervention.

Plasma lipoprotein levels and metabolism can be affected by the type of dietary fats [3]. We now know several aspects of the effects on plasma lipids and lipoproteins of individual monounsaturated (MUFA) and polyunsaturated fatty acids (PUFA), and of their utility in the management of hyperlipidemia.

PUFA and Lipoproteins

Oleic acid (c18:1ω9) is the dominant MUFA in both animal and vegetable fats. Linoleic acid (LA, c18:2ω6) is the most common ω6 PUFA, found in many vegetable oils such as safflower, corn, sunflower seed, and soybean oils. α-linolenic (LNA, c18:3ω3), eicosapentaenoic (EPA, c20:5ω3) and docosahexaenoic (DHA, c22:6ω3) acids are the major ω3 PUFAs and they are present in elevated concentrations in green leafy vegetables, marine foods and fish oils respectively [4].

Many studies have compared the effects of MUFAs and PUFAs on plasma lipoproteins [3]. Mattson and Grundy [5] compared two diets enriched in LA or oleic acid with a diet enriched in saturated fatty acid (palmitic acid). All three diets were liquid-formula diets providing 40% of calories as fat. These authors found that LA and oleic acids have the same low density lipoprotein-cholesterol (LDL-C)-lowering effect, but more interestingly MUFAs did not reduce the high density lipoprotein-cholesterol (HDL-C) levels as usually observed with PUFAs. We have examined the effects on lipid metabolism of a natural product particularly enriched in oleic acid, i.e. olive oil, usually consumed in the Mediterranean countries [6]. We showed a decrease of total and LDL cholesterol by 10 and 12% respectively when using olive oil in comparison to a standard low fat diet. Total HDL-C and HDL subfractions cholesterol levels remained unchanged during the two dietary periods.

The primary mechanism for LDL lowering by both MUFA and PUFA is an increment of the fractional catabolic rate of LDL, while the HDL-lowering effect of LA might be due to a reduction in the HDL synthesis [3].

Patients with hyperlipidemia are at high risk for coronary heart disease, and for this reason they may benefit from a dietary treatment of the lipid disorder. The studies performed in hypercholesterolemic patients suggest that a diet enriched in MUFAs can be safely used in these patients because of the LDL-lowering effect without adverse effects on HDL-C [3]. In contrast, a high amount of ω6 PUFAs can lower HDL-C, while ω3 PUFAs do not change total and LDL-C levels in these patients.

Several studies have examined the effects of diets enriched with unsaturated fatty acids in hypertriglyceridemic patients [3, 7]. A triglyceride-lowering activity of unsaturated, both MUFA and ω6 PUFA, has not been consistently observed. On the other hand, very long chain ω3 PUFAs are very effective in lowering very low density lipoprotein (VLDL)-triglyceride levels. They markedly reduce triglycerides by an inhibition of the hepatic synthesis of VLDL particles. When saturated fats were maintained constant during ω3 fatty acid supplementation, LDL-C levels increased in hypertriglyceridemic patients [7].

Hypertriglyceridemia is still a controversial vascular risk factor [8]. Austin et al. [9] showed that the presence of small and dense LDL particles, with high plasma triglyceride levels, so-called 'lipoprotein pattern B', is associated with cardiovascular disease.

We studied the effects of 2.5 g/day of ω-3 PUFA ethyl esters on lipoprotein levels and LDL size in 17 hypertriglyceridemic patients [10]. Plasma levels fell by 31%, and both cholesterol and triglycerides in the VLDL fraction were reduced after 8 weeks of treatment. LDL size, measured by gradient gel electrophoresis, was smaller in the hypertriglyceridemic patients in comparison to that of healthy controls (25.2 ± 0.4 vs. 27.2 ± 0.8 nm, mean $\pm$ SD. $p < 0.001$). However, the LDL size remained unchanged throughout the study despite the triglyceride-lowering effect of ω3 fatty acids. The LDL diameter was always inversely related to plasma triglyceride levels, while no relationships were found between LDL size and the percent content of EPA, DHA, oleic and LA in plasma phospholipids.

Only a few studies have been performed on the effects of unsaturated fatty acids in patients with combined hyperlipidemia, a common metabolic disorder associated with premature coronary heart disease [3, 7]. As pointed out by Failor et al. [11], the primary difference in response to PUFAs of combined hyperlipidemic patients is the potent triglyceride-lowering effect of ω3 PUFAs in comparison to ω6 PUFAs. However, in these patients ω3 fatty acids usually increase LDL-C levels.

Conclusions

We have today various possibilities to greatly modify lipoprotein levels and composition by using different fatty acids. Based on the available knowledge, the dietary management of hyperlipidemias is a rational and effective approach to prevent atherosclerotic vascular disease.

References

1   Keys A: Coronary heart disease in seven countries. Circulation 1970;41:1–211.
2   Ball MJ: Dietary intervention trials – Effect on cardiovascular morbidity and mortality. Curr Opin Lipidol 1993;3:7–12.
3   Grundy SM, Denke MA: Dietary influences on serum lipids and lipoproteins. J Lipid Res 1990; 31:1149–1172.
4   Goodnight SH Jr, Harris WS, Connor WE, et al: Polyunsaturated fatty acids, hyperlipidemia, and thrombosis. Arteriosclerosis 1982;2:87–113.
5   Mattson FH, Grundy SM: Comparison of effects of dietary saturated, monounsaturated, and polyunsaturated fatty acids on plasma lipids and lipoproteins in man. J Lipid Res 1985;26:194–202.
6   Baggio G, Pagnan A, Muraca M, et al: Olive-oil enriched diet: Effect on serum lipoprotein levels and biliary cholesterol saturation. Am J Clin Nutr 1988;47:960–964.
7   Harris WS: Fish oil and plasma lipid and lipoprotein metabolism in humans: A critical review. J Lipid Res 1989;30:785–807.
8   Austin MA: Plasma triglyceride and coronary heart disease. Arterioscler Thromb 1991;11:2–14.
9   Austin MA, Breslow JL, Hennekens CH, et al: Low-density lipoprotein subclass patterns and risk of myocardial infarction. JAMA 1988;260:1917–1921.
10 Manzato E, Zambon S, Cortella A, et al: Effects of n-3 polyunsaturated fatty acids on lipoprotein physical properties in hypertriglyceridemic patients (abstract). XI Int Symp Drugs Affecting Lipid Metabolism, Firenze 1992.
11 Failor RA, Childs T, Bierman EL: The effects of omega-3 and omega-6 fatty acid-enriched diets on plasma lipoproteins and apoproteins in familial combined hyperlipidemia. Metabolism 1988;37: 1021–1028.

Gaetano Crepaldi, MD, Department of Internal Medicine, Via Giustiniani, 2,
I-35128 Padova (Italy)

Galli C, Simopoulos AP, Tremoli E (eds): Effects of Fatty Acids and Lipids in
Health and Disease. World Rev Nutr Diet. Basel, Karger, 1994, vol 76, pp 41–44

....

Future Directions in ω6 and ω3 Fatty Acid Research

P.J. Nestel

CSIRO Division of Human Nutrition, Adelaide, S.A., Australia

Despite the long-standing recognition of the cholesterol-lowering effect of linoleic acid (LA) several crucial questions remain unanswered: (a) Does LA lower low density lipoprotein (LDL) independently of displacing saturates; (b) the relative potencies of LA versus oleic acid; (c) the optimal amount of LA especially in relation to the ω6/ω3 fatty acid ratio; (d) the mechanism whereby LA lowers plasma LDL levels; (e) the effect of LA on high density lipoprotein (HDL) cholesterol and on plasma triglyceride levels; (f) the effectiveness of other ω6 fatty acids: 18:3 and 20:3.

Two recent reports in table 1 [1, 2] suggest a greater cholesterol lowering by LA than by oleic acid. However, this remains a controversial issue.

Evidence that LA lowers plasma cholesterol independently of displacing saturated fatty acids was provided by us [3] in a study in which LA was added rather than substituted. Despite the resultant high fat intake, LDL cholesterol fell.

Although LA is by far the major ω6 fatty acid in our diet, claims of specific benefits for γ-linolenic acid and its product dihomo-γ-linolenic acid have been made. This is based on the postulate that additional dietary 18:3 and 20:3 ω6 fatty acids will overcome any age or disease-related Δ6-desaturase deficiency.

The effects on other lipids are summarized in table 2 [4–6]. In particular, the effect on HDL is critical yet remains contentious. The evidence shows that HDL is lowered only at high intakes of LA.

Dietary studies in humans and monkeys suggest that the desirable amount of LA in terms of LDL lowering is about 6% energy [7]. Limiting LA intake to this will lead to a dietary ω6/ω3 ratio of about 4, assuming a modest increase in plant foods and fish. The optimum ratio needs to be established, especially with

Table 1. Effects of ω6 PUFA on LDL

Trial	Ref.
Meta-analysis of 27 trials Δ LDL = 1.28 SFA–0.24 MUFA–0.55 PUFA	1
Dietary trial Δ LDL = 2.6 Δ 14:0+2.9 Δ 16:0–0.5 Δ 18:0–0.7 Δ PUFA	2
Conclusion: Effect of PUFA > MUFA	

SFA = Saturated fatty acids; MUFA = monounsaturated fatty acids; PUFA = polyunsaturated fatty acids.

Table 2. Effects of ω6 PUFA on other lipids

Effect	Ref.
Linoleate lowers HDL cholesterol	
Yes	4
No	1, 2, 5
Minor triglyceride-lowering; diminished alimentary lipemia	
Lp(a) – no effect	6

respect to eicosanoid formation. Finally, dietary 18:2 may reduce the LDL-raising effect of dietary cholesterol [6].

Despite decades of research, the LDL-lowering effect of LA is uncertain probably because of its multiplicity of actions. Despite reports of diminished very low density lipoprotein (VLDL) and LDL production, increased LDL removal and increased sterol excretion, none of these has been incontrovertibly proven as the key factor. Recent interest centers around the LDL receptor. Whereas studies in the guinea pig, the hamster and cebus monkey have shown increased LDL receptor-mediated removal, this has not been shown conclusively in man.

Some of the key questions for ω3 fatty acid research are: (a) the biology of the actions of the major ω3 fatty acids on lipid metabolism; (b) the precise reasons for the triglyceride-lowering effect (the major lipid outcome); (c) the unpredictable effects on LDL levels and the reasons for this, and (d) the net benefit of the lipid changes in terms of atherosclerosis.

Table 3. Hepatic triglyceride balance and cholesterol balance

Triglyceride		Cholesterol	
input	output	input	output
↓ Fatty acid synthesis	↑ Fatty acid oxidation	↓ Absorption	↓ Lipoprotein secretion
↓ Esterification (PAP, ADGAT)	↓ VLDL triglyceride	↓ Synthesis ↔	↑ Biliary excretion
↓ Plasma FFA	secretion	? Lipoprotein uptake	↑ CE storage

PAP = Phosphatidate phosphohydrolase; ADGAT = acyl CoA: 1,2 diacylglycerol acyltransferase; FFA = free fatty acid; VLDL = very low density lipoprotein; CE = cholesteryl ester.

Table 4. Issues which still need to be resolved [for review, see 9]

Triglycerides and fatty acids: unresolved issues	Fish oil on LDL metabolism: unresolved questions
Is FFA flux from adipose tissue reduced?	Why and when does LDL concentration rise?
Is increased FFA oxidation peroxisomal?	How is LDL removal affected?
Which esterifying enzymes are lowered?	How is LDL receptor activity regulated?
Is apoB formation and secretion reduced?	Is LDL composition changed?
Are EPA and DHA equipotent?	Is LDL size changed?
	Do EPA and DHA exert similar effects?
	Do 'fish oil' LDL function abnormally?
	Is LDL cholesterol rise atherogenic?

α-Linolenic acid, the most abundant ω3 fatty acid in the human diet and the precursor of the highly potent 20:5 eicosapentaenoic acid (EPA) and 22:6 docosahexaenoic acid (DHA), has surprisingly little effect on lipoproteins [5]. While there is agreement about the triglyceride-lowering and HDL-raising effects of EPA, recent reports from Berge's [8] laboratory have cast doubt on the corresponding efficacy of DHA.

The 'triglyceride balance' and 'cholesterol balance' in the liver can be summarized as in table 3, but issues that still need to be resolved are shown in table 4 [for review, see 9]. In particular there is considerable uncertainty about the nature of the paradoxical effects of fish oil on LDL metabolism. The increase in LDL cholesterol occurs mostly in hyperlipidemic subjects, in whom the small, dense cholesterol-poor LDL may convert to lighter cholesterol-rich LDL. Despite the rise in LDL cholesterol this may be less atherogenic, since it is dense LDL, which are also more prone to oxidation, that are the more atherogenic species.

A major question that needs answering is whether the change in lipoproteins brought about by fish oil is a major factor in the lowering of coronary heart disease. Although HDL cholesterol levels rise, the reduction in VLDL results in small, cholesterol-rich species, while the potential increase in LDL is of uncertain importance.

References

1   Mensink RP, Katan KB: Effect of dietary fatty acids on serum lipids and lipoproteins. Arterioscler Thromb 1992;12:911–919.
2   Derr J, Kris-Etherton PM, Pearson TA, et al: The role of fatty acid saturation on plasma lipids, lipoproteins, and apolipoprotiens. II. The plasma total and low density lipoprotein cholesterol response of individual fatty acids. Metabolism 1993;42:130–134.
3   Rassias G, Kestin M, Nestel PJ: Linoleic acid lowers LDL cholesterol without a proportionate displacement of saturated fatty acid. Eur J Clin Nutr 1991;45:315–320.
4   Mattson FH, Grundy SM: Comparison of effects of dietary saturated, monounsaturated, and polyunsaturated fatty acids on plasma lipids and lipoproteins in man. J Lipid Res 1985;26:194–202.
5   Kestin M, Clifton P, Belling GB, et al: n-3 fatty acids of marine origin lower systolic blood pressure and triglycerides but raise LDL-cholesterol compared with plant n-3 and n-6 fatty acids. Am J Clin Nutr 1990;51:1028–1034.
6   Brown SA, Morrisett J, Patsch JR, et al: Influence of short-term dietary cholesterol and fat on human plasma Lp(a) and LDL levels. J Lipid Res 1991;32:1281–1289.
7   Hayes KC, Khosla P: Dietary fatty acid thresholds and cholesterolemia. FASEB J 1992;6:2600–2607.
8   Willumsen N, Hexeberg S, Skorve J, et al: Docosahexaenoic acid shows no triglyceride-lowering effects but increases the peroxisomal fatty acid oxidation in liver of rats. J Lipid Res 1993;34:13–22.
9   Nestel PJ: Effects of n-3 fatty acids on lipid metabolism. Annu Rev Nutr 1990;10:149–167.

P.J. Nestel, MD, Chief, CSIRO Division of Human Nutrition, PO Box 10041 Gouger Street, Adelaide, SA 5000 (Australia)

Galli C, Simopoulos AP, Tremoli E (eds): Effects of Fatty Acids and Lipids in
Health and Disease. World Rev Nutr Diet. Basel, Karger, 1994, vol 76, pp 45–46

······························

Summary Statement:
ω3 Fatty Acids and Thrombosis

Babette B. Weksler

The session was co-chaired by *B. B. Weksler* and *R. Paoletti,* and presentations were made by Drs. *Weksler, A. Nordøy, S. Endres, E. Tremoli, G. Di Minno,* and *M. B. Donati.*

This session examined the concept that dietary ω3 fatty acids affect hemostatic mechanisms so as to decrease likelihood of thrombosis especially in the setting of vascular disease.

It was pointed out that effects of ω3 fatty acids on hemostasis and thrombosis are multifactorial. Initial studies that concentrated on antithrombotic changes in platelet function have more recently been re-evaluated and show that decreases in platelet aggregation or mediator release are modest even when large amounts of ω3 fatty acids are consumed. What also must be considered are changes in vascular endothelial and monocyte function, such as increases in vascular prostacyclin or nitric oxide production by endothelium, and decreases in tissue factor and mitogen production by monocytes, which have secondary antithrombotic effects. Prolongation of bleeding time by ω3 fatty acids has not been shown to lead to increased surgical bleeding when directly tested. Anti-inflammatory effects of ω3 fatty acids on neutrophils and monocytes may have important implications in decreasing vascular injury and opposing development of atherosclerotic lesions.

It was emphasized that to see antithrombotic effects of dietary ω3 fatty acids in patients at high risk of occlusive cardiovascular disease, it is necessary to decrease dietary saturated fat content as well as add ω3 fatty acids. Recent data indicate that the combination of ω3 fatty acids and lovastatin prevented exercise-induced shortening of bleeding time, indicating a favorable effect upon platelet-vessel wall interaction. This combination also enhanced decreases in plasma triglycerides and factor VII. Elevation of the latter has been associated with high risk of cardiovascular events.

The effects of ω3 fatty acids on cytokine synthesis by monocytes and macrophages were discussed. It was pointed out that the inflammatory effects of cytokines may contribute to the development of atherosclerosis as well as to classic inflammatory diseases, particularly interleukin-1 as an inducer of prothrombotic substances in endothelial cells and in macrophages. Studies in normal subjects and patients with inflammatory disease have shown a marked suppression of interleukin-1 synthesis, probably taking place at the transcriptional level, during treatment with ω3 fatty acids that persists for many weeks after this treatment has been stopped. The long duration of treatment effect is of great interest for the design of treatment regimens.

Studies on human monocyte/macrophages derived from subjects treated with ω3 fatty acid ethyl ester concentrates, examining particularly the inhibitory effect of ω3 treatment on monocyte procoagulant activity, indicated that long-term treatment, of more than 6 weeks, was needed to show an inhibitory effect. Moreover, more prolonged treatment was required for the effect to persist. With 18 weeks of treatment, monocyte functions remained depressed for > 6 months, even though the monocyte lipid concentrations had returned to normal. These findings were further supported by data on the long-lasting inhibition of platelet response to collagen following ω3 fatty acid ethyl ester administration, and by the associated finding of decreased von Willebrand factor binding to thrombin-stimulated platelets that occurred with long-term use of ω3 fatty acids which persisted after ω3 discontinuation, even though thromboxane production had returned to normal. A persistence of elevated DHA in platelet membranes was considered to be an important key to this long-term change in platelet function resulting from ω3 fatty acid administration. A persistence of DHA in membrane lipids was described elsewhere in the meeting.

The new 'GISSI Prevention' study was discussed. The study, which will begin soon, has as its major aim the reduction of risk of atherosclerotic cardiovascular events in patients who have already had one myocardial infarction. Four groups of subjects will be included, randomized in addition to dietary advice, to ω3 fatty acids alone, vitamin E alone, the combination of ω3 fatty acids and vitamin E, or dietary advice only. For those with high cholesterol, a second randomization is planned after 6 months of first intervention, to add pravastatin or placebo. This large trial will be the first of its type in scope, duration, and use of ω3 fatty acids in cardiovascular prophylaxis of patients at high risk of occlusive events.

Galli C, Simopoulos AP, Tremoli E (eds): Effects of Fatty Acids and Lipids in
Health and Disease. World Rev Nutr Diet. Basel, Karger, 1994, vol 76, pp 47–50

....

ω3 Fatty Acids Have Multiple Antithrombotic Effects

Babette B. Weksler

Division of Hematology/Oncology, Department of Medicine, The New York Hospital–
Cornell Medical Center, New York, N.Y., USA

Both the physiologic process of hemostasis and its pathologic equivalent, thrombosis, are influenced by dietary lipids. The addition to a diet of ω3 polyunsaturated fatty acids (PUFA) or their substitution for more saturated lipid components alters the function of many different elements in hemostasis. What is not yet established is whether or not these changes have antithrombotic effects or if they promote bleeding.

Attention was first drawn to possible antithrombotic effects of ω3 PUFA by the observation that Greenland Eskimos consuming a diet rich in cold-water fish, had long bleeding times and a low rate of myocardial infarction [1]. Fish-eating has since been epidemiologically associated with low rates of myocardial infarction in other countries as well, such as Japan, The Netherlands and England [2–4].

Platelet Effects of ω3 Fatty Acids

Because of the prolonged bleeding times of Greenland Eskimos, studies of hemostatic effects of fish, fish oils, or ω3 PUFA derived from fish oil initially were focused on platelet function. Confirmation of prolonged bleeding time, reduced serum thromboxane and decreased platelet aggregation was obtained in response to eating fish (mackerel or salmon), fish oils or purified ω3 PUFA [5, 6]. The effects were roughly dose- and time-dependent. These observations were made on normal subjects as well as on patients with hyperlipidemia or atherosclerosis. One study in hypercholesterolemic patients showed that shortened platelet survival returned toward normal during ω3 PUFA therapy, presumably resulting from decreased in vivo platelet activation [7]. Recent studies using concentrated ethyl esters of ω3 PUFA as dietary supplements show similar results, with more variable effects on platelet function, raising the question of equivalency of effects of fish/fish oils and more concentrated ω3 fatty acid esters. Incorporation of docosahexaenoic acid (DHA) into platelet membranes may have long-lasting and different effects from those of eicosapentaenoic acid (EPA).

To evaluate the question if ω3 PUFA supplementation might increase bleeding risk, patients undergoing coronary artery bypass surgery were treated for 1 month preoperatively with 4.3 g/day of EPA and DHA, during which a significant decrease in platelet aggregation, a fall in serum thromboxane production, and a moderate increase in bleeding time were documented [8]. At surgery, however, no increase in intraoperative or postoperative bleeding, nor increased need for transfusion, was observed compared to control subjects not treated with ω3 PUFA [8].

Indeed, in a recent study of very high dose ω3 PUFA treatment of baboons (1 g/kg/day) in a platelet-dependent thrombosis model, depression of platelet aggregation and prolongation of bleeding time were not more profound than in humans treated at a much lower dosage ω3 PUFA (0.06–0.01 g/kg/day) [9]. This indicated that a very high dose of ω3 PUFA did not inhibit hemostasis more than a low dose. However, in these baboons, platelet adhesion to vascular shunts or to arteries injured by angioplasty was strikingly decreased. As a result, less thrombosis occurred and there was less intimal thickening following the vascular injury. These findings suggest (1) that ω3 PUFA may produce effects quantitatively more antithrombotic than antihemostatic, and (2) that actions of ω3 PUFA on the vessel wall rather than on platelets directly may contribute to a net antithrombotic effect.

Effects of ω3 Fatty Acids on the Vessel Wall

Further confirmation of the concept that ω3 PUFA may be antithrombotic by acting on the vascular wall or on blood elements other than platelets comes first from studies of vascular wall function and second, from studies of leukocyte function.

In the coronary artery bypass patients treated with ω3 PUFA prior to surgery, direct examination of aortic, saphenous vein and atrial tissue was made to determine if changes in vascular prostaglandin production occurred. Significant increases in prostacyclin (PGI$_2$) production capacity of these vascular tissues, representing arteries with atherosclerotic change (aorta), normal vein (saphenous vein) and microvasculature (atrial appendage) were observed

[8]. Other studies have indicated that ω3 PUFA may also enhance vascular production of nitric oxide, the endothelial-derived relaxing factor (EDRF) that also inhibits platelet activation [10]. Production by endothelium of growth factors for smooth muscle cell proliferation is inhibited by ω3 PUFA [11]. Moreover, in the baboon study (see above), ω3 PUFA-treated vascular segments were significantly less adherent for platelets [9].

Effects of ω3 Fatty Acids on Leukocyte Function

A related area of interest in ω3 PUFA involves the control of inflammation. Studies both in normal subjects and patients with inflammatory joint disease such as rheumatoid arthritis have indicated that ω3 PUFA therapy results in many decreased activities of neutrophils and monocytes, including chemotaxis, adherence to protein substrates, superoxide formation, and production of leukotrienes, cytokines, mitogens, tissue factor, and procoagulant activity [12, 13]. Since inflammatory changes in the vascular wall produced by cytokines, oxygen radicals, or infiltrating leukocytes can render the vascular surface prothrombotic, the effects of ω3 PUFA on leukocytes have long-range antithrombotic potential. As pointed out in recent studies, these anti-inflammatory effects of ω3 PUFA, especially on monocyte function, may last for weeks after cessation of therapy [13], although their onset may be slow and thus not perceived during short-term treatment.

Finally, ω3 PUFA treatment reduces serum triglyceride levels and has variable effects upon circulating fibrinogen, plasma procoagulants and components of the fibrinolytic system [14]. At the present time, effects on plasma proteins appear to be of lesser physiologic importance compared to the effects on vascular wall, leukocytes and platelets.

References

1   Dyerberg J, Bang HO, Stofferson E, et al: Eicosapentaenoic acid and prevention of thrombosis and atherosclerosis. Lancet 1978;ii:117–119.
2   Hirai A, Hamazaki T, Terano T: Eicosapentaenoic acid and platelet function in Japanese. Lancet 1980;ii:1132–1133.
3   Kromhout D, Bosschieter EB, Coulander CL: The inverse relation between fish consumption and 20-year mortality from coronary heart disease. N Engl J Med 1985;312:1205–1216.
4   Burr ML, Gilbert JF, Holliday RM, et al: Effects of changes in fat, fish and fibre intakes on death and myocardial infarction: Diet and Reinfarction Trial (DART). Lancet 1989;ii:757–761.
5   Von Schacky C, Fischer S, Weber PC: Long-term effects of dietary marine omega-3 fatty acids upon plasma and cellular lipids, platelet function and eicosanoid formation in humans. J Clin Invest 1985;76:1626–1631.
6   Goodnight SH, Harris WS, Connor WE: The effect of dietary omega-3 fatty acids on platelet composition and function in man: A prospective, controlled study. Blood 1981;58:880–885.

7   Levine PH, Fisher M, Schneider PS, et al: Dietary supplementation with omega-3 fatty acids prolongs platelet survival in hyperlipidemic patients with atherosclerosis. Arch Intern Med 1989; 149:1113–1117.

8   DeCaterina R, Giannessi D, Mazzone A, et al: Vascular prostacyclin is increased in patients ingesting omega-3 polyunsaturated fatty acids before coronary artery bypass graft surgery. Circulation 1990;82:428–438.

9   Harker LA, Kelly AB, Hanson SR, et al: Interruption of vascular thrombus formation and vascular lesion formation by dietary n-3 fatty acids in fish oil in nonhuman primates. Circulation 1993;87: 1017–1029.

10 Shimokawa H, Vanhoutte PM: Dietary omega-3 fatty acids and endothelium-dependent relaxation in porcine coronary arteries. Am J Physiol 1989;256:H968–H973.

11 Fox PL, DiCorleto PE: Fish oils inhibit endothelial cell production of platelet-derived growth factor-like protein. Science 1988;241:453–456.

12 Lee TN, Hoover RL, Williams JD, et al: Effect of dietary enrichment with eicosapentaenoic and docosahexaenoic acids on in vitro neutrophil and monocyte leukotriene generation and neutrophil function. N Engl J Med 1985;312:1217–1224.

13 Endres S, Ghorbani R, Kelley VE, et al: The effect of dietary supplementation with n-3 polyunsaturated fatty acids on the synthesis of interleukin-1 and tumor necrosis factor by mononuclear cells. N Engl J Med 1989;320:265–271.

14 Goodnight SH: Mechanism of the antithrombotic effects of fish oil. Clin Hematol 1990;3:601–623.

Babette B. Weksler, MD, Professor of Medicine, Division of Hematology/Oncology,
Room C-608, Cornell University Medical College, 1300 York Avenue, New York, NY 10021 (USA)

Galli C, Simopoulos AP, Tremoli E (eds): Effects of Fatty Acids and Lipids in
Health and Disease. World Rev Nutr Diet. Basel, Karger, 1994, vol 76, pp 51–54

....

ω3 Fatty Acids and Cardiovascular Risk Factors

Arne Nordøy, John-Bjarne Hansen

Department of Medicine, Institute of Clinical Medicine, University of Tromsø, Norway

A series of risk factors associated with increased incidence of premature cardiovascular disease have been defined in epidemiological studies and confirmed in prospective intervention studies and in experimental investigations. The most important risk factors are a high intake of dietary saturated fats, hypercholesterolemia, combined hyperlipemia, smoking, hypertension, diabetes and familial occurrence of premature cardiovascular disease [1]. However, only about 50% of the mortality caused by coronary heart disease (CHD) may be accounted for by these established risk factors. In recent years risk factors related to the occurrence of arterial thrombosis have been established [2]. These risk factors include changes in platelet number and platelet function, increased coagulation activities related to fibrinogen and factor VII, and reduced fibrinolytic activity. A combination of metabolic disturbances have been included in the 'metabolic syndrome' or the 'atherothrombogenic syndrome' [3]. This cluster of established risk factors including insulin resistance, central obesity, hypertriglyceridemia, low high density lipoprotein (HDL) cholesterol level and raised blood pressure may have particular relevance for the occurrence of CHD in women. Homocystinemia and hyperferritinemia have also been suggested to represent independent risk factors for CHD.

The ω3 fatty acids (FA), eicosapentaenoic acid (EPA) and docosahexaenoic acid (DHA) from fish and fish products have in epidemiological and experimental studies been associated with a low mortality of CHD and a reduced tendency to development of atherosclerotic and thrombotic lesions [for review, see 4]. A series of physiological consequences of the intake of ω3 FA have been linked to the beneficial effects of these FA. Some have been directly associated with the effects on established risk factors.

Dietary Saturated Fat

Both in Eskimos [5] and other population groups with a high intake of $\omega3$ FA and a low occurrence of CHD, the diet has also been low in saturated FA. In other population groups, particularly in Northern Europe, a high intake of $\omega3$ FA has been combined with a high intake also of saturated FA [1]. In these groups the incidence of CHD is high. Recent studies may indicate that a high intake of saturated FA may abolish the beneficial effects of dietary $\omega3$ FA on thrombogenic risk factors [6]. In addition, when the intake of saturated FA is high, the level of low density lipoprotein (LDL) cholesterol also remains high after supplement of $\omega3$ FA. Thus, the atherogenic and thrombogenic potentials of saturated FA may disturb some of the beneficial effects of $\omega3$ FA.

Hypercholesterolemia

$\omega3$ FA seem to be without significant effects on the level of LDL in patients with familial hypercholesterolemia. However, the effect of $\omega3$ FA on platelet vessel wall interaction reflected in bleeding time and other platelet function tests and in the effect on the myocardial cells may indicate that $\omega3$ FA also have a potential in the optimal treatment of patients with this metabolic disease [7].

Other Hyperlipidemias

The most constant and striking effects of $\omega3$ FA have been a reduction of very low density lipoprotein (VLDL) and total triglycerides [4]. In women, hypertriglyceridemia has been established as an independent risk factor for development of CHD [1]. In recent years, hypertriglyceridemia has also been considered a risk factor for CHD in men; however, this hypertriglyceridemia has usually been associated with a probably more important reduction of HDL. Furthermore, these combined lipid disturbances are often associated with other risk factors. In patients with such metabolic abnormalities the effect of $\omega3$ FA on blood lipids and blood pressure may be of great significance.

Smoking

Smoking has been established as an independent risk factor for CHD. The mechanisms by which smoking influences the processes involved have not been commonly agreed upon. Recent studies have confirmed that smoking increases the level of oxidized LDL in plasma. This may relate to the harmful effects of smoking. Increased intake of the highly polyunsaturated ω3 FA may further increase the tendency to oxidation of LDL. Recent studies give no consistent answer to whether supplements of ω3 FA to the diet further increase oxidized products above that induced by smoking alone [8, 9].

Hypertension

In subjects with a moderate hypertension the dietary supplement of ω3 FA significantly reduces both systolic and diastolic blood pressure [4]. In preliminary studies we have shown that the effects more likely are related to DHA rather than to EPA [Bønaa et al., unpubl observations]. Many studies have confirmed the effects of ω3 FA on blood pressure and within the next few years it is predicted that ω3 FA may be recommended as a dietary supplement to most patients with hypertension. Certainly, this also includes patients where hypertension is associated with other metabolic disturbances which may all increase the tendency to atherothrombogenic disease.

Thrombogenic Factors

In recent years many factors directly associated with thrombogenesis have been suggested as independent risk factors. They include changes in platelets, like increased platelet number, increased platelet volume and increased aggregability; changes in coagulation with high levels of fibrinogen and factor VII, and decreased fibrinolytic activity associated particularly with increased levels of inhibitor of tissue plasminogen activator (PAI-1). Also, other changes like high white cell counts and high hematocrit levels have been suggested as independent risk factors. Obviously, at present we are just at the beginning of an area where new risk factor profiles will give us the possibility to further increase the reliability with regard to predicting the occurrence of CHD. The prolonged bleeding time, the reduced platelet aggregability and the reduced production of thrombogenic eicosanoids like thromboxane A_2 (TXA_2) in the Eskimos initiated our understanding of the association between ω3 FA and these risk factors. In recent intervention studies the effects of ω3 FA in subjects on a Western diet have only given modest effects on platelet function, except for a rather consistent reduction in platelet TXA_2 production. The reduction in serum triglycerides induced by ω3 FA has been associated with a reduction in factor VII whereas reports on the effects on fibrinolysis have been conflicting. Some investigators have reported an increase in PAI-1 which has not been confirmed by others.

In summary, the beneficial effects of the very long chain polyunsaturated fatty acids EPA and DHA from fish oil have been documented on many of the established risk factors, and on other physiological processes which may be equally important for the prevention of CHD. The effects so far are mainly related to blood lipids and platelet vessel wall interaction. The effects of ω3 FA are influenced by the content of other dietary FA, in particular the saturated FA and the ω6 FA. Studies under way investigating the effects of the individual FA from fish oils may increase the rationale for the use of these FA in clinical medicine.

References

1   Nordøy A, Goodnight SH: Dietary lipids and thrombosis: Relationships to atherosclerosis. Arteriosclerosis 1990;10:149–163.
2   Nordøy A: Haemostatic factors in coronary heart disease. J Intern Med 1993;233:377–383.
3   Reaven GM: Role of insulin resistance in human disease. Diabetes 1988;37:1595–1607.
4   Nordøy A: Is there a rational use for n-3 fatty acids (fish oils) in clinical medicine? Drugs 1991; 42:331–342.
5   Bang HO, Dyerberg J, Sinclair HM: The composition of the Eskimo food in Northwestern Greenland. Am J Clin Nutr 1980;33:2657–2666.
6   Nordøy A, Hatcher L, Goodnight S, et al: Effects of dietary fatty acids and fish oil upon eicosanoid production and haemostatic parameters in normal men. Lab Clin Med 1994;in press.
7   Hansen JB, Lyngmo V, Svensson B, et al: Inhibition of exercise-induced shortening of the bleeding time by fish oil in familial hypercholesterolemia. Arterioscler Thromb 1993;13:98–104.
8   Harats D, Daback Y, Hollander G, et al: Fish oil ingestion in smokers and nonsmokers enhances peroxidation of plasma lipoproteins. Atherosclerosis 1991;90:127–139.
9   Alessandrini P, Cagzolato G, Soldan S, et al: Treatment of type IIB hyper-lipidemic patients with concentrated n-3 fatty acids does not result in increased peroxidation of LDL (abstract). Eur Atheroscler Soc 61st Meet, Capri. Naples, Ariello Bros Press, 1993.

Arne Nordøy, MD, PhD, Department of Medicine, Institute of Clinical Medicine, University of Tromsø, N-9038 Tromsø (Norway)

Galli C, Simopoulos AP, Tremoli E (eds): Effects of Fatty Acid and Lipids in
Health and Disease. World Rev Nutr Diet. Basel, Karger, 1994, vol 76, pp 55–59

....

Effects of ω3 Fatty Acid Ethyl Esters on Monocyte Tissue Factor Expression

*Elena Tremoli, Sonia Eligini, Susanna Colli, Paola Maderna,
Franca Marangoni, Maria Teresa Angeli, Cesare R. Sirtori, Claudio Galli*

Institute of Pharmacological Sciences, E. Grossi Paoletti Center,
University of Milan, Italy

ω3 polyunsaturated fatty acids derived from fish or fish oils have been extensively studied in the last years for their potential antithrombotic and antiatherosclerotic effects [1, 2]. Several studies indicate that, among other effects on blood cells and on components of the vessel wall, ω3 fatty acids influence various functions of monocytes, i.e. chemotaxis [3], cytokine synthesis [4] and leukotriene generation [5] and the expression of mRNA for platelet-derived growth factor [6]. Dietary administration of fish oil to nonhuman primates has been recently demonstrated to impair hemostatic function, including tissue factor expression by endotoxin-stimulated mononuclear cells [7]. Tissue factor (TF) is a membrane-bound glycoprotein, present in a variety of cells, including monocytes and endothelial cells, that activates blood coagulation.

In the present study the potential effects of ω3 fatty acid ethyl esters, administered to healthy volunteers, on tissue factor expression by adherent monocytes were investigated.

Subjects and Methods

Sixteen healthy subjects (8 males and 8 females, age range 23–39 years), were selected. The subjects were randomly assigned to treatments A or B. *Treatment A:* 3 g/day (3 capsules, 1 g each, containing ω3 fatty acid ethyl esters) plus 3 g/day (3 capsules, 1 g each) olive oil. *Treatment B:* 6 g/day of fatty acid ethyl esters. After 6 weeks of treatment all the subjects followed a 12-week period with 3 g of ω3 fatty acid ethyl esters/day.

Monocyte Isolation and Stimulation

Blood, anticoagulated with 3.8% sodium citrate (9:1, v/v), was centrifuged at 150 g for 18 min to obtain platelet-rich plasma, that was discarded. The residue was processed for mononuclear cell isolation using centrifugation on Ficoll-Paque as previously described [8]. Monocytes were purified by adhesion to plastic. For TF activity (TFa) determination, adherent monocytes were incubated for 4 h at 37 °C in 5% CO_2 humid atmosphere in the absence and in the presence of 10 µg/ml LPS. At the end of the incubation, cells were scraped off and subjected to three cycles of freezing and thawing.

Lipids from monocyte membranes were extracted with chloroform/methanol (2:1 v/v) containing 5 µg/ml of BHT. The contents of the extracts were determined by microgravimetry. Total phospholipids were isolated from lipid extracts by TLC, using hexane/diethyl ether/acetic acid 80:20:1 v/v/v, as developing agents. The zones containing total phospholipids were scraped off. Fatty acid methyl esters were prepared by transmethylation using methanolic HCl. The methyl esters were separated by gas liquid chromatography on capillary columns (Supelcowax 10, fused silica 30 m, 0.32 ID, µm film) using programming temperature (140–210 °C at 2.5/min increments).

Assay of Tissue Factor Activity

Total cellular content of TFa was determined by one-stage clotting assay at 37 °C on disrupted monocytes. Assay mixture contained 0.1 ml of cell lysates, 0.1 ml citrated pooled normal plasma and 0.1 ml $CaCl_2$ 25 mM. Results were expressed in arbitrary units (U) by comparison to a standard curve of clotting times, obtained by serial dilutions of human placental thromboplastin.

Results

Adherent monocytes isolated from subjects in treatments A and B expressed 7.0 ± 1.1 and 6.5 ± 0.8 units/µg protein TFa respectively (n = 8) and stimulation of the cells with 10 µg/ml LPS induced an almost twofold increase in TFa in both groups of subjects. Six-week treatment with either 3 or 6 g/day of ω3 fatty acid ethyl esters did not influence basal and LPS-stimulated expression of TFa (fig. 1). The treatment with ω3 fatty acid ethyl esters for 6 weeks resulted in a dose-dependent accumulation of eicosapentaenoic acid (EPA) in plasma lipids, whereas docosahexaenoic acid (DHA) accumulated in the two groups to a comparable extent (fig. 2). After 12 weeks of treatments with ω3 fatty acid ethyl esters a dramatic impairment in the expression of TFa was observed in unstimulated and LPS-stimulated monocytes of subjects assigned to both treatments (fig. 1). A further reduction of monocyte TFa was observed in both groups compared with baseline also at 18 weeks after the start of the treatments. Levels of EPA and DHA were similarly increased in both groups of subjects after 12 and 18 weeks of treatments.

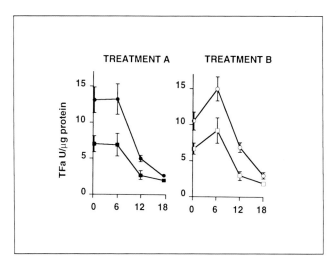

Fig. 1. TFa expression by monocytes during treatment A (3 g/day of ω3 fatty acid ethyl esters for 18 weeks) and treatment B (6 g/day ω3 fatty acid ethyl esters for 6 weeks followed by 12 weeks with 3 g/day). Each point represents the average ± SEM units TFa/μg protein for 8 subjects in each group at each time interval. □,■ = Unstimulated monocytes; ○,● = LPS-stimulated monocytes. Values in both treatment A and B at 12 and 18 weeks of treatment were significantly different ($p < 0.001$) from values at baseline and at 6 weeks of treatment in both unstimulated and LPS stimulated cells.

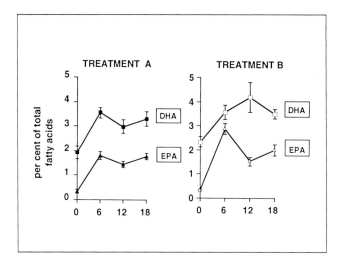

Fig. 2. Levels of ω3 fatty acid in monocytes expressed as percent of total fatty acids during treatments. $p < 0.001$ versus baseline for all comparisons.

Discussion

These data indicate that the administration of moderate doses of ω3 fatty acid ethyl esters to healthy subjects reduces TFa in unstimulated monocytes as well as in cells stimulated with LPS. Inhibition of TFa in human mononuclear cells following fish oil administration to healthy subjects was previously reported by Hansen et al. [9]. This observation, however, was not confirmed by others [10, 11]. It is worth mentioning that these latter studies were performed with relatively high doses of ω3 fatty acids (around 5 g/day) administered for relatively short time periods (4–8 weeks). Recently, it has been shown that the administration of 1 g/kg of an ω3 ethyl ester concentrate to nonhuman primates for 12 weeks dramatically impairs monocyte TFa [7]. The delay in the onset of the effects of ω3 fatty acid ethyl esters on monocyte TFa suggests that TFa regulation is not an immediate target for the fatty acids but that they probably act in concert with other, yet unidentified mediators, possibly influencing the gene expression. This hypothesis is consistent with the finding that ω3 fatty acid ethyl esters inhibit gene expression of platelet-derived growth factor A and B in monocytes [6]. The inhibition of TFa was accompanied by a consistent accumulation of EPA and DHA in plasma and in monocyte compartments. During treatments, however, we have not found a strict correlation between TFa inhibition and the levels of EPA or DHA in plasma and/or monocyte lipids. One should consider, however, that fatty acid analyses were performed on total lipids, possibly masking discrete changes in the different phospholipid pools. Fatty acid composition of specific monocyte intracellular compartments might represent a more accurate marker to be correlated with changes in TFa.

In conclusion, the results reported in this study indicate that ω3 fatty acid ethyl esters may profoundly influence the generation of thrombin affecting the capacity of monocytes to express procoagulant activity.

References

1 Nordøy A, Goodnight S: Dietary lipids and thrombosis. Arteriosclerosis 1990;10:149–163.
2 Leaf A, Weber PC: Cardiovascular effects of ω3 fatty acids. N Engl J Med 1988;318:549–557.
3 Schmidt EB, Pedersen JO, Ekelund S, et al: Cod liver oil inhibits neutrophil and monocyte chemotaxis in healthy males. Atherosclerosis 1989;77:53–57.
4 Endres SR, Ghorbani VE, Kelley VE, et al: The effects of dietary supplementation with n-3 polyunsaturated fatty acids on the synthesis of interleukin-1 and tumor necrosis factor by mononuclear cells. N Engl J Med 1989;320:265–271.
5 Lee TH, Hoover RL, Williams JD, et al: Effect of dietary enrichment with eicosapentaenoic and docosahexaenoic acids on in vitro neutrophil and monocyte leukotriene generation and neutrophil function. N Engl J Med 1985;312:1217–1224.
6 Kaminski WE, Jendraschak E, Kiefl R, et al: Dietary ω3 fatty acids lower levels of platelet-derived growth factor mRNA in mononuclear cells. Blood 1993;81:1871–1879.

7   Harker LA, Kelly AB, Hanson SR, et al: Interruption of vascular thrombus formation and vascular lesion formation by dietary n-3 fatty acids in fish oil in nonhuman primates. Circulation 1993; 87:1017–1029.

8   Tremoli E, Stragliotto E, Colli S, et al: Effects of n-3 fatty acids on monocyte function; in De Caterina R, Kristensen SD, Schmidt EB (eds): Fish Oil and Vascular Disease. Current Topics in Cardiovascular Disease. New York, Springer, 1993, pp 79–84.

9   Hansen JB, Olsen JO, Wilsgard L, et al: Effects of supplementation with cod liver oil on monocyte thromboplastin synthesis, coagulation and fibrinolysis. J Intern Med 1989;225:133–139.

10 Muller AD, Van Houwelingen AC, van dam Mieras MCE, et al: Effect of moderate fish intake on haemostatic parameters in healthy males. Thromb Haemost 1989;61:468–473.

11 Pellegrini G, Totani L, Di Santo A, et al: Supplementation-induced changes in polyunsaturated fatty acid membrane and plasma composition do not modify mononuclear cell procoagulant activity. Thromb Res 1993;71:95–101.

Prof. Elena Tremoli, Pharmacology of Thrombosis and Atherosclerosis Laboratory,
Institute of Pharmacological Sciences, University of Milan, Via Balzaretti, 9,
I-20133 Milan (Italy)

Galli C, Simopoulos AP, Tremoli E (eds): Effects of Fatty Acids and Lipids in
Health and Disease. World Rev Nutr Diet. Basel, Karger, 1994, vol 76, pp 60–63

....

Platelet Effects of ω3 Fatty Acid Ethyl Esters

Ferdinando Cirillo[a], *Antonio Coppola*[a], *Umberto Piemontino*[a],
Vincenzo Marottoli[b], *Aldo Amoriello*[c], *Anna Maria Cerbone*[a],
Edoardo Stragliotto[d], *Elena Tremoli*[e], *Giovanni Di Minno*[a],
Mario Mancini[a]

[a] Clinica Medica, Institute of Internal Medicine and Dismetabolic Diseases,
   University of Naples 'Federico II', Naples;
   Divisions of [b]Hematology and [c]Immunohematology, Cardarelli Hospital, Naples;
[d] Medical Department, Farmitalia-Carlo Erba, Milan;
[e] Institute of Pharmacological Sciences, University of Milan, Italy

Over the last 10 years, in parallel with studies showing that fish consumption was associated with a lower tendency to major ischemic complications of
atherosclerosis [1], a series of studies has investigated some mechanisms
through which fish oils may exert their antithrombotic potential. The studies
indicate [2–7] that fish oils lower plasma fibrinogen, cholesterol and triglyceride
levels, and affect the synthesis of prostaglandins, thromboxanes and leukotrienes, and the release of growth factors. As far as their antiplatelet effect is
concerned, the majority of the ex vivo studies [8–11] show that the supplementation of these fats is associated with a prolonged bleeding time and with
changes in the ability of platelets to aggregate and to adhere to collagen or
fibrinogen-coated surfaces. Pathophysiological and clinical data [12] suggest
that the aggregation of platelets involves to the same extent the synthesis of
prostaglandins/thromboxane. On the other hand, the skin bleeding time and
platelet adhesion appear to be largely independent of the formation of these
arachidonic acid (AA) metabolites. Recently, Tremoli et al. [pers. commun.]
have shown that, following cessation of the administration of a prolonged
regimen of ω3 fatty acid ethyl esters, a long-lasting inhibition of platelet
aggregation takes place that is associated with major changes in platelet
membrane lipid composition, and that little involves the synthesis of prosta-

glandins and thromboxane. We reasoned that this could be a major direction to elucidate potential antiplatelet mechanisms of ω3 fatty acids independent of the synthesis of prostaglandins and thromboxane. We also thought that it could be clinically important to evaluate whether a long-lasting platelet inhibitory effect of ω3 fatty acids could be detected after a relatively short-course supplementation of ω3 fatty acids.

Methods

The study was carried out in 10 healthy volunteers (8 m, 2 f; 24–30 years old), who were instructed to ingest twice daily, during the meals, 2.55 g of ω3 fatty acid ethyl esters (eicosapentaenoic (EPA)/docosahexaenoic acid (DHA) ratio: 1.4) for 4 weeks. Immediately before starting the supplementation, 28 days later (i.e. on the day when it was withdrawn) and monthly for 3 months after the cessation of the supplementation, the skin bleeding time was carried out, and blood platelets were analyzed for their membrane glycoprotein content, their ability to adhere to glass, to aggregate in response to collagen and ADP, to form thromboxane B_2 (TXB$_2$), to secrete ATP and to bind radiolabelled fibrinogen in response to thrombin and to form cAMP under basal conditions and following incubation with 1 mM PGE$_1$. The skin bleeding time; platelet adhesion to glass; platelet aggregation in response to threshold concentrations of ADP, collagen and AA; fibrinogen purification, characterization and labelling; TXB$_2$ measurements; platelet cAMP; platelet secretion of nucleotides and binding to platelets of radiolabelled fibrinogen was carried out as previously reported [12, 13]. Fluorescence-activated cell sorter (FACS) analysis of membrane glycoproteins was carried out as recently reported [14].

Results and Discussion

At the cessation of the supplementation, the bleeding time was slightly although not significantly prolonged, while platelet adhesion to glass was unchanged. Both these measurements were unchanged in the 3 months following cessation of ω3 fatty acid supplementation. In contrast, aggregation in response to collagen was greatly affected. At the cessation of the supplementation, the amount of collagen needed to cause 50% aggregation within 3 min was almost twice as much as that required before starting the supplementation (p < 0.01). The effect on the aggregation of platelets was detectable only after a prolonged lag phase, the significant differences reported above being detectable 1 and 2 months after withdrawing the supplementation. Samples obtained 1 and 2 months after withdrawing ω3 fatty acids were comparable with respect to these platelet defects (p > 0.05). Similar results were found when ADP was the agonist employed. In contrast, the response to AA was unchanged throughout the study. Together, the data were taken to suggest that the inhibitory effect observed was at least in part independent of the synthesis of thromboxane as well as the effect of ω3 fatty acids on the prostaglandin endoperoxide/thromboxane receptor on the platelet surface. On the other hand, platelet secretion of ATP was normal, as were platelet levels of cAMP. Thus other directions were explored. The binding of fibrinogen to specific platelet receptors is a prerequisite for the aggregation of platelets [12]. Therefore, we wondered whether the impaired aggregation of platelets that follows cessation of the ω3 fatty acid supplementation could involve an abnormal binding of this protein to platelets, and the data obtained showed that this was not the case. On the other hand, FACS analysis showed that the glycoprotein IIb-IIIa complex, the receptor for fibrinogen on platelet surface, was normal in cells from these normal volunteers. Thus the impaired aggregation of subjects who had ingested ω3 fatty acids did not appear to involve TXB_2 formation, nucleotide secretion, platelet cAMP or abnormalities of cell surface glycoprotein IIb-IIIa. On the other hand, ω3 FA did not appear to affect the glycoprotein Ib, the glycoprotein Ib-IX complex, and the glycoprotein receptors for thrombospondin, PADGEM and vLAβ.

Recent data from Tremoli et al. suggest that the long-lasting impaired aggregation of platelets that follows the cessation of a prolonged ω3 fatty acid supplementation is associated with changes in the platelet membrane lipid composition. The association between the two phenomena is now under intensive investigation in our laboratory and is supported by in vitro studies [15] in which enrichment of platelets with EPA and/or DHA was associated with aggregation defects comparable to those of our volunteers. In conclusion, we have found that following cessation of supplementation of ω3 fatty acid ethyl esters a long-lasting inhibition of platelet aggregation occurs that is independent of the binding of fibrinogen to its specific receptor.

References

1   Nordoy A, Goodnight SH: Dietary lipids and thrombosis. Relationship to atherosclerosis. Arteriosclerosis 1990;10:149–163.
2   Sanders TAB: Influence of n-3 fatty acids on blood lipids; in Simopoulos AP, Kifer RR, Martin RE, Barlow S (eds): Health Effects of ω3 Polyunsaturated Fatty Acids in Seafoods. World Rev Nutr Diet. Basel, Karger, 1991, vol 66, pp 358–366.
3   Nestel PJ, Connor WR, Reardon MF, et al: Suppression by diets rich in fish oil of very low density lipoprotein production in man. J Clin Invest 1984;74:82–89.
4   De Caterina R, Giannessi D, Mazzone A, et al: Vascular prostacyclin is increased in patients ingesting n-3 polyunsaturated fatty acids before coronary artery bypass graft surgery. Circulation 1990;82:428–438.
5   Lee TH, Hoover RL, Williams JD, et al: Effect of dietary enrichment with eicosapentaenoic and docosahexaenoic acids on in vitro neutrophil and monocyte leukotriene generation and neutrophil function. N Engl J Med 1985;312:1217–1224.
6   Smith DL, Willis AI, Nguyen N, et al: Eskimo plasma constituents dihomo-gamma-linolenic acid, eicosapentaenoic acid and docosahexaenoic acid inhibit the release of atherogenic mitogens. Lipids 1989;24:70–75.

7   Schmidt EB, Varming K, Ernst E, et al: Dose-response studies on the effect of n-3 polyunsaturated fatty acids on lipids and haemostasis. Thromb Haemost 1990;63:1–5.

8   Thorngren M, Gustafson A: Effects of 11-week increase in dietary eicosapentaenoic acid on bleeding time, lipids and platelet aggregation. Lancet 1981;ii:1190–1193.

9   Hirai A, Terano T, Hamazaki T, et al: The effects of the oral administration of fish oil concentrate on the release and the metabolism of ^{14}C arachidonic acid and ^{14}C-eicosapentaenoic acid by human platelets. Thromb Res 1982;28:285–298.

10 Von Schacky C, Fisher S, Weber PC: Long-term effects of dietary marine n-3 fatty acids upon plasma and cellular lipids, platelet function, and eicosanoid formation in humans. J Clin Invest 1985;76:1626–1631.

11 Li X, Steiner M: Fish oil: A potent inhibitor of platelet adhesiveness. Blood 1990;76:938–945.

12 Di Minno G, Cerbone AM, Mattioli PL, et al: Functionally thrombasthenic state in normal platelets following the administration of ticlopidine. J Clin Invest 1985;75:328–338.

13 Di Minno G, Silver MJ, de Gaetano G: Prostaglandins as inhibitors of human platelet aggregation. Br J Haematol 1979;43:637–647.

14 Stragliotto E, Camera M, Postiglione A, et al: Functionally abnormal monocytes in hypercholesterolemia. Artherioscler Thromb 1993;13:944–950.

15 Croset M, Lagarde M: In vitro incorporation and metabolism of icosapentaenoic and docosahexaenoic acids in human platelets. Effect on aggregation. Thromb Haemost 1986;56:57–62.

Giovanni Di Minno, Clinica Medica, Institute of Internal Medicine and Metabolic Diseases, University of Naples 'Federico II', Via S. Pansini, 5, I–80131 Napoli (Italy)

Galli C, Simopoulos AP, Tremoli E (eds): Effects of Fatty Acids and Lipids in
Health and Disease. World Rev Nutr Diet. Basel, Karger, 1994, vol 76, pp 64–65

....

Summary Statement:
Fatty Acids and Cancer

David Horrobin

The session was co-chaired by *D.F. Horrobin* and *G. Noseda*, and presentations were made by Drs. *Horrobin, C. Borek, W.T. Cave, Jr., K.C.H. Fearon,
A.S. Spector,* and *M.J. Tisdale.*

The relevance of unsaturated lipids in cancer and the possible use of such
lipids in cancer is a rapidly developing topic. The basic observations made on
cultured cells were reviewed in this session which have led to the current
interest. Two of the research groups have found that polyunsaturated fatty
acids are cytotoxic to malignant cells while having little or no adverse action
against normal cells. One group found that highly unsaturated ω3 fatty acids
can also make leukemic cells more susceptible to attack by hyperthermia or
cytotoxic drugs and the effects are associated with iron-stimulated lipid peroxidation. Parinaric acid, containing conjugated double bonds, was a particularly
effective cytotoxic agent. It was also reported that prooxidants could enhance,
and antioxidants inhibit, the anticancer effects of essential fatty acids (EFA)
and that both γ-linolenic acid (GLA) and eicosapentaenoic acid (EPA) slowed
or blocked the growth of human malignant melanoma or breast cancer trans-
planted into nude mice.

The interactions between EPA and docosahexaenoic acid (DHA) and
ionizing radiation and oncogene transfection were reported. Both EPA and
DHA could block the formation of transformed foci produced by either route.
The protective effects were associated with changes in the composition of
phospholipids, particularly involving the replacement of arachidonic acid (AA)
with the ω3 fatty acids.

The relationship between neoplasia and isoprenoid synthesis was discussed. Various regulatory proteins, including the *ras* oncogene product p-21,
require isoprenylation to express their effects. This offers the possibility of

regulating oncogene activity by inhibiting isoprenoid synthesis, for example, by inhibition of early stages of cholesterol synthesis using drugs such as lovastatin. This reduces growth of malignant cells both in vivo and in vitro.

Extensive work was described using the MAC-16 colonic adenocarcinoma and the production of cachexia. This tumor is not responsive to cytotoxic drugs. EPA was able both to reduce tumor growth and to suppress the cachexia caused by the tumor. The effect was particularly related to enhanced cell loss rather than to any reduction in tumor growth. The clinical implications of this work were reviewed with particular reference to pancreatic cancer. Both GLA and EPA are able to suppress growth of pancreatic cancer cells in vitro. The initial findings of a phase II clinical trial of lithium-GLA in patients with inoperable pancreatic cancer was described. The drug was well tolerated with a clear improvement in immune function and a possible prolongation of survival.

The overall impression of this session was that unsaturated lipids play important roles in the biology of cancer and may prove to be of therapeutic importance.

Galli C, Simopoulos AP, Tremoli E (eds): Effects of Fatty Acids and Lipids in
Health and Disease. World Rev Nutr Diet. Basel, Karger, 1994, vol 76, pp 66–69

....

ω3 Fatty Acids as Anticarcinogens: Cellular and Molecular Studies

Carmia Borek

Department of Physiology, Tufts University School of Medicine,
Boston, Mass., USA

The Role of Fatty Acids in Cancer

Epidemiological and experimental studies provide compelling evidence that dietary fat modulates the incidence of certain types of cancer [1]. The data show that the quality and composition of fat, not merely its amount, determines the nature of the effects [1–4].

Diets rich in ω6 fatty acids increase the incidence of tumors in colon, pancreas and mammary gland, of animals treated with carcinogens [2]. By contrast, animals fed fish oil diets, that are rich in ω3 fatty acids, especially eicosapentaenoic acid (20:5ω3, EPA) and docosahexaenoic acid (22:6ω3, DHA), and are low in ω6 fatty acids, show a reduced frequency of these tumors [2]. Diets high in fish oil inhibit ultraviolet (UV) light-induced skin tumors in mice [3].

Studies on the effects of ω6 and ω3 fatty acids on human cancer are limited [1]. Eskimos who consume diets rich in fat from marine animals, show a low incidence of cancer [4]. Intervention trials show a negative correlation between total ω3/total ω6 fatty acids and a risk for cancer mortality [1].

The molecular mechanisms which underlie the effects of polyunsaturated fatty acids on cancer development are poorly understood. The complexity of homeostasis in animals makes it virtually impossible to glean the direct effect of specific dietary fatty acids on neoplasia at a molecular level. The utilization of fatty acids and production of metabolites vary from tissue to tissue, and a plethora of effects on cellular membranes and signalling pathways obscure many events that may take place.

Effects of ω3 and ω6 Fatty Acids on Transformation in vitro

Cell transformation in vitro [5] provides a powerful tool to study the effects of lipids in multistep carcinogenesis [6]. The model affords the opportunity to investigate, under defined conditions, the effects of specific ω3 and ω6 fatty acids on neoplastic transformation and shed light on molecular mechanisms that underlie the effect.

Experiments were designed to study dose-related and time-related effects of EPA, DHA and arachidonic acid (AA) on oncogenic transformation by 4 Gy of γ-rays or by a mutated Harvey *ras* (H-*ras*) oncogene, in mouse C_3H $10T\frac{1}{2}$ cells [7]; to examine the effect of these polyunsaturated fatty acids on tumor promotion by TPA in radiation-initiated cells [6], and to examine the capacity of these fatty acids to modulate prostaglandin production in the cells.

Experiments were also designed to analyze the effect of ω3 fatty acids on the molecular species of four major classes of phospholipids in cells growing under the same conditions as those in the studies on transformation [7].

ω3 Fatty Acids Inhibit Transformation by Radiation

The results showed that EPA and DHA inhibited radiogenic transformation by 80–100%. AA had no inhibitory effects and in some cases enhanced transformation frequency by 50% [7]. To achieve maximum protection against transformation EPA must be present continuously from the period before radiation treatment or just after. The presence of EPA, during the initiation phase of transformation [6], i.e. in the first 2 weeks after exposure to radiation, in a 6-week experiment, was critical for effective inhibition of transformation.

ω3 Fatty Acids Inhibit H-ras Oncogene-Induced Transformation

EPA and DHA suppressed transformation by the H-*ras* oncogene. The effects were dose-related [7]. Maximum suppression was observed at $100 \, \mu M$ with a lesser effect at $50 \, \mu M$ and more at $25 \, \mu M$. AA had no inhibitory effects. Maximum protection was conferred by EPA and DHA when the fatty acids were present continuously following transfection. When EPA was withdrawn after 2 weeks the inhibitor effect was lost [7].

ω3 Fatty Acids Inhibit Promotion

EPA completely blocked the promoting effects of TPA on radiation-initiated cells. One interpretation of this effect may be that since EPA suppressed initiation by radiation, it reduced the number of cells which are susceptible to the promoting effect of the phorbol ester. Another interpretation may be that since radiation and TPA activate protein kinase C [6], EPA may be acting by attenuating the activation of protein kinase C and affecting transformation and promotion, as has been observed with other bioactive lipids [6].

ω3 Fatty Acids Inhibit Prostaglandin Production

Both EPA and DHA reduced the capacity of C_3H $10T\frac{1}{2}$ cells to produce PGE_2, suggesting that these ω3 fatty acids may inhibit transformation by impairing the formation of cyclooxygenase-derived products. Earlier studies showing that radiation at 4 Gy induces the release of AA and increases the production of PGE_2, in C_3H $10T\frac{1}{2}$ cells [Borek, unpubl.], support the possibility that ω3 fatty acids inhibit transformation by antagonizing the AA cascade.

ω3 Fatty Acids After Phospholipid Composition

Growth of cells in EPA or DHA resulted in an extensive remodelling of the molecular species in phosphatidylcholine, phosphatidylethanolamine, phosphatidylserine and phosphatidylinositol, so that ω3 fatty acids-containing species replaced ω6 fatty acids-containing species [7]. A shift to phospholipids containing high levels of ω3 fatty acids could modify the affinity of membrane receptors compared to membranes containing predominantly ω6 fatty acids. In this manner the affinity of cellular receptors that play a role in transformation could be affected by the incorporation of ω3 fatty acids, thereby inhibiting transformation. Unesterified ω3 fatty acids in the membrane containing reduced AA-containing species could also compete with unesterified AA for oxygenases. This could also interfere with transformation.

Conclusion

Cancer is a multistep process. It is initiated by genetic damage [5, 8] and progresses through a series of promotional steps during which deregulated cell growth takes place. Transformation represents a progressive disorder in signal transduction pathways which regulate gene expression. Cellular phospholipids and the AA cascade play an important role in signalling events and in transformation. Exposure to EPA and DHA produces a remodelling of cellular phospholipids so that they are substantially reduced in AA-containing species. Such changes would reduce eicosanoid synthesis, modulate signal transduction pathways and attenuate the activation of protein kinase C by shifting diacylglycerol species rich in AA, which constitute the endogenous ligand for protein kinase C to species rich in ω3 fatty acids. The action of transforming agents which may act in part via activation of protein kinase C, such as radiation, the H-*ras* oncogene or the phorbol ester tumor promoter TPA [6], may be inhibited and transformation suppressed. Remodelling of phospholipids may affect growth factor and hormone receptors, and intercellular communication, all of which may play a role in the transformation process [9].

References

1   Galli C, Butrum R: Dietary ω3 fatty acids and cancer: An overview; in Simopoulos et al (eds): Health Effects of ω3 Polyunsaturated Fatty Acids in Seafoods. World Rev Nutr Diet. Basel, Karger, 1991, vol 66, pp 446–461.
2   Belury MH, Locniskar M, Fischer SM: Modulation of phorbol-ester associated events in epidermal cells by linoleate and arachidonate. Lipids 1993;28:407–413.
3   Orengo IF, Black HS, Kettler AH, Wolf JE Jr: Influences of dietary menhaden oil upon carcinogenesis and various cutaneous responses to ultraviolet radiation. Photochem Photobiol 1989;49:71–77.
4   Bang HO, Dyerberg J, Hijorne N: The composition of food consumed by Greenland eskimos. Acta Med Scand 1976;200:69–73.
5   Borek C, Ong A, Mason H: Distinctive transforming genes in X-ray transformed mammalian cells. Proc Natl Acad Sci USA 1987;84:97–98.
6   Borek C, Ong A, Stevens VL, Wang E, Merril AH Jr: Long chain (sphingoid) bases inhibit multistage carcinogenesis in mouse C_3H 10T$\frac{1}{2}$ cells treated with radiation and phorbol myristate 13-acetate. Proc Natl Acad Sci USA 1991;88:1953–1957.
7   Takahashi M, Porzetakiewicz M, Ong A, Borek C, Lowenstein JM: Effect of ω3 and ω6 fatty acids on transformation of cultured cells by irradiation and transfection. Cancer Res 1992;52:154–162.
8   Motokura T, Arnold A: Cyclins and oncogenes. Biochim Biophys Acta 1993;1155:63–78.
9   Borek C: The induction and regulation of radiogenic transformation in vitro; in Grunberger D, Goff S (eds): Mechanisms of Cellular Transformation by Carcinogenic Agents. New York, Pergamon Press, 1987, pp 151–195.

Carmia Borek, Department of Physiology, Tufts University School of Medicine, Boston, MA 02111 (USA)

Galli C, Simopoulos AP, Tremoli E (eds): Effects of Fatty Acids and Lipids in
Health and Disease. World Rev Nutr Diet. Basel, Karger, 1994, vol 76, pp 70–73

....

Isoprenoids and Neoplastic Growth

William T. Cave, Jr.

Department of Medicine, University of Rochester School of Medicine,
Rochester, N.Y., USA

While the importance of posttranslational modification of proteins has
been well understood for a long time, it has only been very recently recognized
that isoprenoids have an important role in this process [1–4]. Experimental
studies on a wide variety of mammalian cells have now confirmed that many
important intracellular regulatory proteins are prenylated, and that inhibition
of this process critically affects cell function.

Biochemistry

Prenylation is an enzymatic process involving the covalent attachment of
either a 15-carbon (farnesyl), or 20-carbon (geranylgeranyl) isoprenoid moiety
to a cysteine at the carboxyl-terminal end of the substrate protein. There
appears to be one farnesyl transferase enzyme and at least two geranylgeranyl
transferases. Specificity results from the ability of the different prenyl transfer-
ases to selectively recognize unique amino acid motifs at the COOH terminus
of individual proteins. In mammalian cells, more than 80% of all protein-
bound isoprenoid is in the geranylgeranyl form.

Physiological Importance

The functional value of prenylation is hypothesized to be its ability to
provide certain cytosolic proteins with a hydrophobic membrane anchor. The

observation that all of the prenylated proteins are at least partially localized to cell membranes supports this concept; however, the fact that some have both cytosolic and membrane-bound locations suggests that prenylation alone, while necessary, may not be a sufficient condition for membrane association. There is also some evidence to indicate that certain prenylated proteins may have a direct regulatory function in cell metabolism and growth, in addition to participating in the processing and assembly of macromolecular structures.

The recent discovery that many members of the *ras* protein superfamily undergo prenylation has emphasized the potential importance this process may have in such critical intracellular functions as vesicular transport, cytoskeletal organization, and phagocytosis. The prenyl modification appears to allow these proteins to cycle between distinct membrane compartments, and between membrane-associated and cytosolic states. Other prenylated proteins (e.g. the γ subunits of the heterotrimeric G proteins, and rhodopsin kinase) serve as integral elements in signal transduction pathways. Through a GTP(on) GDP(off) switch, these G proteins allow extracellularly oriented receptors to link with intracellular membrane-bound effectors. Collectively, these data argue that prenylation, probably through several mechanisms, greatly influences both intercellular localization and biological activity of a number of important molecules.

Role in Oncogene Product Function

The three human *ras* genes (H-, K-, and N-*ras*) encode four structurally related proteins which function as molecular switches through their interactions with guanidine nucleotides. Oncogenic *ras* proteins containing single amino acid substitutions at residues 12, 13, or 61 are defective in GDP/GTP cycling, and persist constitutively in the active, GTP-bound form. As such, they chronically activate growth stimulatory pathway(s) and promote tumorigenesis. These proteins are initially synthesized as soluble cytoplasmic proteins, lacking any conventional transmembrane or hydrophobic sequences, and must then be post translationally processed in order to become attached to cell membranes. This processing initially involves the attachment of a farnesyl group to a cysteine, four amino acids from the carboxy-terminus of the molecule, by a thioether bond, then the proteolytic removal of the three terminal amino acid residues, the carboxymethylation of the now terminal farnesyl-cysteine residue, and finally, the palmitylation of one of the cysteines upstream from the carboxy-terminus [5]. The subsequent association of the *ras* proteins with the inner face of the plasma membrane is a critical requirement for their function.

Once prenylation was recognized to be essential for oncogenic *ras* protein function, it also became apparent that its inhibition could have potential chemotherapeutic value. In this regard, several different pharmacological approaches have been explored. One approach has been to inhibit mevalonate synthesis (a precursor step to isoprenoid formation) by using the HMG-CoA reductase inhibitors lovastatin or compactin. A number of investigators have shown that these agents can arrest the in vitro growth of a variety of neoplastic cells which express the *ras* oncogene, such as the T24 bladder carcinoma cell line [6] and the UR61 rat pheochromocytoma cell line [7]. It has also been noted that this effect is not due to a reduction in the cellular content of the *ras* oncoprotein (detected immunocytochemically), but rather to the inability of the oncoprotein to attach to the cell membrane. Some concern, however, has been expressed about the potential practicality of using these agents in cancer therapy, because protein prenylation may be less sensitive to inhibition than other pathways that arise from mevalonate biosynthesis. Nevertheless, such global inhibition could be useful if certain tumor cells are found to have an enhanced susceptibility to mevalonate inhibition relative to their normal counterparts [8]. In vivo experiments reporting that lovastatin therapy can diminish NMU-induced mammary tumor growth in rats [9] and human bladder tumor H-*ras* oncogene transformed 3T3 cell tumors in nude mice [10] seem particularly relevant in this regard.

Another pharmacological approach has been to block *ras* oncoprotein isoprenylation with agents that selectively inhibit individual prenyl transfer-ases. Certain synthetic peptides, for example, containing the COOH-terminal amino acid motif of the *ras* oncoproteins, have proven effective in vitro and in vivo inhibitors of *ras* prenylation. Problems regarding their introduction into cells and their delivery to the appropriate intracellular targets, however, have currently limited their clinical usefulness. A somewhat more promising alternative, so far, has been the use of inhibitory molecules such as limolene, and its more potent derivatives perillic acid and dihydroperillic acid. These monoterpenes have been shown to selectively interfere with the isoprenylation of the *ras* oncoprotein [11], and delay the growth of two different types of carcinogen-induced rat mammary tumors.

Conclusion

The recent discovery that prenylation is an important posttranslational modification required by a number of regulatory proteins has extended our insight into how lipids and fatty acids influence growth and development. Already the recognition of its pivotal role in the activation of specific oncogenic

---

proteins has created exciting opportunities for new therapeutic approaches to cancer therapy. Certainly, as our biochemical understanding of this process further improves, other therapeutic opportunities will emerge.

References

1 Maltese WA: Posttranslational modification of proteins by isoprenoids in mammalian cells. FASEB J 1990;4:3319–3328.
2 Khosravi-Far R, Cox AD, Kato K, et al: Protein prenylation: Key to *ras* function and cancer intervention? Cell Growth Differ 1992;3:461–469.
3 Rine J, Kim S-H: A role for isoprenoid lipids in the localization and function of an oncoprotein. New Biol 1990;2:219–226.
4 Sinensky M, Lutz R: The prenylation of proteins. Bioessays 1992;14:25–30.
5 Casey PJ, Solski PA, Der CJ, et al: P21*ras* is modified by a farnesyl isoprenoid. Proc Natl Acad Sci USA 1989;86:8323–8327.
6 Jakobisiak M, Bruno S, Skierski JS, et al: Cell cycle specific effects of lovastatin. Proc Natl Acad Sci USA 1991;88:3628–3632.
7 Mendola CE, Backer JM: Lovastatin blocks n-*ras* oncogene-induced neuronal differentiation. Cell Growth Differ 1990;1:499–502.
8 Kato K, Der CJ, Buss JE: Prenoids and palmitate: Lipids that control the biological activity of *ras* proteins. Cancer Biol 1992;3:179–188.
9 Cave WT: 3-Hydroxy-3-methylglutaryl-coenzyme A (HMG-CoA) reductase inhibitor effects on mammary tumor development in rats fed high fat diets. Proc. Am Assoc Cancer Res 1992;33:131.
10 Sebti SM, Tkalcevic GT, Jani JP: Lovastatin, a cholesterol biosynthesis inhibitor, inhibits the growth of human H-*ras* oncogene-transformed cells in nude mice. Cancer Commun 1991;3:141–147.
11 Crowell PL, Lin S, Vedjs E, et al: Identification of metabolites of the antitumor agent *d*-limonene capable of inhibiting protein isoprenylation and cell growth. Cancer Chemother Pharmacol 1992; 31:205–212.

William T. Cave, Jr., MD, Department of Medicine, University of Rochester School of Medicine, Rochester, NY 14642 (USA)

Galli C, Simopoulos AP, Tremoli E (eds): Effects of Fatty Acids and Lipids in
Health and Disease. World Rev Nutr Diet. Basel, Karger, 1994, vol 76, pp 74–76

....

Polyunsaturated Fatty Acids in the Treatment of Weight-Losing Patients with Pancreatic Cancer [1]

J.S. Falconer, K.C.H. Fearon, J.A. Ross, D.C. Carter

University Department of Surgery, Royal Infirmary of Edinburgh, UK

At present there is no effective treatment for patients with unresectable
pancreatic cancer [1]. Recently, polyunsaturated fatty acids (PUFA) such as
eicosapentaenoic acid (EPA) and γ-linolenic acid (GLA) have been shown to
have a selective antineoplastic action both in vitro [2] and in animal models [3].
Moreover, these PUFA are also of interest in pancreatic cancer because they
may normalize the immunosuppression and enhanced cytokine secretion [4]
that is thought to contribute to cachexia in such patients [5].

We have undertaken a phase I clinical trial in pancreatic cancer patients
using the lithium salt of GLA. This report consists of the first 18 patients with
respect to their tolerance of GLA, the effects on T-cell function and proinflammatory cytokine production (tumor necrosis factor, TNF) and the overall
survival of the patients.

Study Protocol

Patients with unresectable pancreatic cancer and who had been endoscopically stented or
undergone surgical bypass (n = 18) received a 10-day continuous intravenous infusion of
GLA. To prevent thrombophlebitis the infusion was given via a central venous catheter and

[1] This work has been supported from grants from the Cancer Research Campaign, the
Melville Trust for the Care and Cure of Cancer and Scotia Pharmaceuticals.

Table 1. Red cell and plasma phospholipid fatty acid concentration before (0) and after treatment for 1 month with GLA: results expressed as mg/100 mg fatty acid (n = 10 patients)

Fatty acid	Plasma		RBC	
	0	1 month	0	1 month
Palmitic acid	30.1 (0.6)	29.4 (0.7)	32.1 (2.2)	28.9 (2.8)
Oleic acid	14.1 (0.5)	11.7 (0.4)*	22.0 (1.5)	19.2 (2.0)
Linoleic acid	19.9 (1.1)	14.3 (1.9)*	7.6 (0.8)	6.7 (0.8)
Dihomo-GLA	3.2 (0.4)	10.1 (1.3)*	0.8 (0.1)	3.2 (0.6)*
Arachidonic acid	10.8 (0.6)	13.8 (0.9)*	9.8 (2.2)	14.0 (2.4)
EPA	0.9 (0.1)	0.5 (0.1)*	0.3 (0.1)	0.3 (0.1)

Mean (SEM); * $p < 0.01$ vs. pretreatment level Student's paired t test.

10,000 units of heparin was added daily. The dose was increased over the first 5 days and then continued at the maximum tolerated dose for the subsequent 5 days. We aimed to achieve a maximum dose of 10 g/day of GLA. Following the period of intravenous GLA, patients received oral GLA to a maximum of 6 g/day. Immunological and cytokine assessments were carried out prior to treatment (day 0), at day 5, day 10 and day 30. At each assessment, peripheral blood mononuclear cells (PBMC) were isolated by differential centrifugation. Thereafter the following assessments were carried out: (1) anti-CD3-stimulated PBMC-tritiated thymidine uptake ex vivo and (2) spontaneous PBMC TNF production ex vivo.

Blood was also taken at time 0 and at 1 month for plasma and red cell phospholipid profiles. Finally the duration of patient survival was also noted.

Results

During the infusion period the mean dose achieved over the second 5 days of the infusion was 5.7 g/day GLA. Out of the 18 patients, 13 progressed onto oral capsules and received a mean dose of 3 g/day. Patient tolerance was good.

The only significant change in red cell phospholipids after 1 month of treatment was a highly significant increase in dihomo-GLA, the immediate metabolite of GLA (table 1). In contrast, there was a significant decrease in plasma oleic acid, linoleic acid and EPA whereas arachidonic acid and dihomo-GLA increased in concentration ($p < 0.01$, Student's paired t test).

Anti-CD3-stimulated T-cell function was significantly increased after 1 month of GLA therapy (29.031 ± 8.223 vs. 58.271 ± 14.625 cpm; day 0 vs. day 30; mean $\pm$ SEM; $p < 0.02$). PBMC TNF production was significantly reduced 1 month after commencement of GLA therapy (981 ± 221 vs. 151 ± 52 pg/ml TNF; $p < 0.01$). To date, the median survival is 8 months (range 1.5–18 months) and 4 patients are still alive.

Conclusion

GLA therapy in pancreatic cancer patients is well tolerated and results in improved T-cell function and reduced pro-inflammatory cytokine production measured ex vivo. The normal median survival of patients with pancreatic cancer is 3–6 months after diagnosis. The patients in this study were treated soon after diagnosis and therefore the median survival of 8 months does not suggest that the survival of pancreatic cancer patients will be markedly prolonged with GLA therapy. It remains to be determined whether administration of GLA via a different protocol or along with cytotoxic drugs might lead to a significant improvement in survival.

References

1   Williamson RCN: Pancreatic cancer: The greatest oncological challenge. Br Med J 1992;296:445–446.
2   Bagen ME, Ellis G, Das UN, Horrobin FD: Differential killing of human carcinoma cells by N3 and N6 polyunsaturated fatty acids. J Natl Cancer Inst 1986;77:1053–1062.
3   Beck SA, Smith KL, Tisdale MJ: Anticachectic and antitumour effect of eicosapentanoic acid and its effect on protein turnover. Cancer Res 1991;51:6089–6093.
4   Endres S, Ghorbani E, Kelley VE, et al: The effect of dietary supplementation with N-3 polyunsaturated fatty acids in the synthesis of interleukin-1 and tumour necrosis factor by mononuclear cells. N Engl J Med 1989;320:265–271.
5   Falconer JS, Fearon KCH, Plester CE, Ross JA, Carter DC: Cytokines the acute phase response and resting energy expenditure in weight-losing pancreatic cancer patients. Ann Surg, in press.

K.C.H. Fearon, University Department of Surgery, Royal Infirmary of Edinburgh,
Lauriston Place, Edinburgh EH3 9YW (UK)

Galli C, Simopoulos AP, Tremoli E (eds): Effects of Fatty Acids and Lipids in
Health and Disease. World Rev Nutr Diet. Basel, Karger, 1994, vol 76, pp 77–80

....

Unsaturated Lipids: A New Approach to the Treatment of Cancer

D.F. Horrobin

Efamol Research Institute, Kentville, N.S., Canada

Several different types of malignant cells have been shown to contain reduced levels of unsaturated lipids [1]. The fatty acids affected are particularly the metabolites of linoleic acid and α-linolenic acid. The picture is consistent either with a defect in Δ-6-desaturation, or with excessive consumption of the metabolites with the normal rate of Δ-6-desaturation being unable to compensate for the loss [1]. Consistent with the deficit of highly polyunsaturated lipids is the fact that malignant cells are also resistant to lipid peroxidation [1, 2]. Even when stimulated by iron, malignant cells generate far smaller quantities of oxidized lipids than do normal cells.

Many existing anticancer agents appear to work in part by generating free radicals which initiate the peroxidation of unsaturated lipids. Unfortunately this process occurs in both normal and malignant cells. The effects on the cancer kill the cancer cells and are therapeutic, but the effects on normal cells are responsible for many of the side effects of cytotoxic therapy. An ideal cancer treatment would be one which could generate free radicals and lipid peroxidation in cancer cells without having the same effects in normal cells. Unsaturated fatty acids with 3, 4 or 5 double bonds appear to be agents in this class.

In vitro Studies

Many studies have now been performed in which highly unsaturated fatty acids are added to normal cells and to malignant cells. The results from several different laboratories have been remarkably consistent [3–8]. Fatty acids with 3, 4 or 5 double bonds consistently kill cancer cells at concentrations which do not harm normal cells. Fatty acids with 2 and 6 double bonds have similar actions but are consistently less potent. In order to observe this phenomenon, depending on the cell line, the culture must be kept going for 5–10 days. Shorter incubations for 24, 48 or 72 h, such as are often employed in screening tests for cytotoxic agents, result in no clear effects.

To date, effects of unsaturated lipids on around 40 different malignant cell lines and 15 different normal cell lines have been reported in the literature. As far as I am aware, all cancer cell lines tested under appropriate conditions have been found to be sensitive to one or more of γ-linolenic acid (GLA), dihomo-γ-GLA (DGLA), eicosapentaenoic acid (EPA) and arachidonic acid (AA). Particularly GLA and DGLA, and to a lesser extent EPA, are nontoxic to normal cells at concentrations which kill cancer cells. For most human cancers, GLA and/or DGLA have been found to be the most effective selective agents, but for colorectal cancers EPA seems to be the most effective.

The cancer cell-killing side effects are accompanied by generation of superoxide and the formation of lipid peroxidation products [1, 9, 10]. Although normal cells exhibit higher levels of background lipid peroxidation than do cancer cells, in the presence of the unsaturated lipids the normal cells show little change in lipid peroxidation whereas the malignant cells show a dramatic increase. There is a good correlation between cell death and the generation of lipid peroxidation. Vitamin E, selenium and other antioxidants can reduce or block the anticancer cell effects of the unsaturated lipids, whereas iron and copper which promote lipid peroxidation can dramatically enhance the cell killing [1].

Multiple passages of cancer cells through concentration of GLA which are below the threshold of cell killing do not seem to introduce any substantial degree of resistance to cytotoxic concentrations. The threshold may be raised by 10–30%, but not more: the cells remain susceptible to the GLA.

In vivo Animal Studies

GLA and EPA have both been tested by administering them orally in the forms of evening primrose and fish oils in a variety of animal tumors, particularly those of the mammary glands. Both GLA and EPA have been shown to inhibit the growth of such tumors [11–14].

Evening primrose and fish oils have also been tested for their ability to inhibit implantation and growth after administration of human breast and melanoma cancers into nude mice. Both oils were approximately equally effective in controlling growth of melanoma, whereas the GLA oil was more effective in blocking the growth of the breast cancer. Both oils also inhibited the implantation of the tumors.

These animal studies encouraged the development of a clinical research program.

Human Studies

Various routes and forms of administration of GLA have been tried in attempts to treat human cancers. High doses of oral GLA in the form of evening primrose oil have produced evidence of responses and prolongation of life without side effects in patients with liver, breast, brain and esophageal cancer [15]. A low dose had no effect in colorectal cancer [16].

While there were no important side effects, patients did have difficulty in swallowing the large volumes of the natural oils and so purified GLA has been developed for oral, intratumoral and intravenous administration. In malignant melanoma the response rate to oral GLA appears to be as good as the response to conventional cytotoxics but without the side effects [R.E. Mansel, pers. commun.]. An excellent and dramatic response has also been obtained in a patient with aggressive radiation-resistant brain cancer. Das [17] has administered GLA into the tumor cavity in a series of patients who have been surgically treated for glioblastoma (a brain cancer which can be palliated but not cured by surgery). He has reported excellent responses in this difficult to manage cancer.

The most promising approach, however, seems to be to administer high doses of the lipids intravenously over a 10-day period, so imitating the conditions of the cell culture experiments. The lithium salts of the EFAs are appropriate vehicles for this. They are partially water soluble and so can be administered intravenously in ordinary aqueous infusion. The only important side effect is hemolysis which can occur if the infusion rate is too rapid and results from a detergent-like dissolution of red cell membranes. However, this is rapidly reversible on stopping the infusion. With experience we have found that it can be prevented by controlling the rate of infusion in each individual patient by monitoring plasma lithium levels. Using ion-selective electrodes this can be done at the bedside in less than 5 min using blood from a finger-prick.

To date, almost 50 patients have been treated, most of them with either breast cancer or pancreatic cancer and most at the Universities of Edinburgh, Cambridge and Cape Town. Initial results in both cancers have been encouraging with suggestive evidence of a dose-related increase in survival.

Conclusion

The use of unsaturated lipids in the treatment of cancer has progressed over 10 years from initial in vitro experiments to human clinical trials. The initial results of these trials indicate that these lipids are likely to find a role as an entirely new approach to the treatment of cancer which involves far less toxicity than has traditionally been associated with this situation.

References

1 Horrobin DF: Essential fatty acids, lipid peroxidation and cancer; in Horrobin DF (ed): Omega-6 Essential Fatty Acids. New York, Wiley-Liss, 1990, pp 351–377.
2 Utsumi K, Yamamoto G, Inaba K: Failure of ferrous iron-induced lipid peroxidation and swelling in the mitochondria isolated from ascites tumor cells. Biochim Biophys Acta 1965;105:368–371.
3 Dippenaar N, Booyens F, Fabri D, et al: The reversibility of cancer: Evidence that malignancy in human hepatoma cells is gamma-linolenic acid deficiency dependent. S Afr Med J 1982;62:683–685.
4 Begin ME, Ells G, Das UN, Horrobin DF: Differential killing of human carcinoma cells supplemented with n-3 and n-6 polyunsaturated fatty acids. J Natl Cancer Inst 1986;77:1053–1062.
5 Begin ME, Ells G, Horrobin DF: Polyunsaturated fatty acid-induced cytotoxicity against tumor cells and its relationship to lipid peroxidation. J Natl Cancer Inst 1988;80:188–194.
6 Fujiwara F, Todo S, Imashuku S: Antitumor effect of gamma-linolenic acid on cultured human neuroblastoma cells. Prostaglandins Leukot Med 1984;15:15–34.
7 Begin MR, Das UN, Ells G: Cytotoxic effects of essential fatty acids in mixed cultures of normal and malignant human cells. Prog Lipid Res 1986;25:573–576.
8 Dippenaar N, Booyens N, Fabri D, et al: The reversibility of cancer: Evidence that malignancy in malanoma cells is gamma-linolenic acid deficiency dependent. S Afr Med J 1982:62:505–509.
9 Das UN, Begin ME, Ells G: Polyunsaturated fatty acids augment free radical generation in tumor cells in vitro. Biochim Biophys Res Commun 1987;145:15–24.
10 Das UN, Huang YS, Begin ME, et al: Uptake and distribution of cis-unsaturated fatty acids and their effects on free radical generation. Free Radic Biol Med 1987;3:9–14.
11 Ghayur T, Horrobin DF: Effects of essential fatty acids in the form of evening primrose oil on the growth of the rat R3230AC transplantable mammary tumor. IRCS J Med Sci 1981;9:582.
12 Pritchard GA, Jones DL, Mansel RE: Lipids in breast carcinogenesis. Br J Surg 1989;76:1069–1073.
13 Karmali RA, Marsh J, Fuchs K: Effects of dietary enrichment with gamma-linolenic acid upon growth of the R3230AC mammary adenocarcinoma. J Nutr Growth Cancer 1985;2:41–51.
14 Gonzales MJ, Schemmel RA, Gray JI: Effect of dietary fat on growth of MCF-9 and MDA-MB231 human breast carcinomas in athymic nude mice: Relationship between tumor growth and lipid peroxidation product levels. Carcinogenesis 1991;12:1231–1235.
15 Van der Merwe CF, Booyens J, Katzoff IE: Oral gamma-linolenic acid in 21 patients with untreatable malignancy. Br J Clin Pract 1987;41:907–915.
16 McIllmurray M, Turkie W: Controlled trial of gamma-linolenic acid in Duke's colorectal cancer. Br Med J 1987;294:1260.
17 Das UN: Gammalinolenic acid, arachidonic acid and eicosapentaenoic acid as potential anti-cancer drugs. Nutrition 1990;6:429–434.

D.F. Horrobin, Efamol Research Institute, PO Box 818, Kentville, NS B4N 4H8 (Canada)

Galli C, Simopoulos AP, Tremoli E (eds): Effects of Fatty Acids and Lipids in
Health and Disease. World Rev Nutr Diet. Basel, Karger, 1994, vol 76, pp 81–85

....

Possible Applications of Polyunsaturated Fatty Acids in Cancer Therapy[1]

Arthur A. Spector[a], *James A. North*[a], *Albert S. Cornelius*[b],
Garry R. Buettner[c]

Departments of [a]Biochemistry, [b]Pediatrics, and the [c]Electron Spin Resonance Facility,
College of Medicine, University of Iowa, Iowa City, Iowa, USA

Humans and animals cannot synthesize the two main forms of polyunsaturated fatty acid (PUFA), the ω6 and ω3 classes. All of these PUFA present in the tissues, including cancer cells, are ultimately derived from the diet. Therefore, the PUFA content and composition of a tumor should be subject to modification, depending on the type and amount of these fatty acids available to the cells. PUFA are important components of cell membranes and are precursors or components of bioactive lipids that have regulatory functions. The dependence of PUFA content on external supply suggests that any property or function of a tumor that is related to PUFA should be susceptible to modification. Based on these considerations, we have explored the potential usefulness of PUFA enrichment or substitution in cancer therapy.

Fatty Acid Compositional Changes

Our initial studies to test the effects of PUFA were done with murine ascites tumors. This work demonstrated that the PUFA content and composition of Ehrlich carcinoma and L1210 leukemia cells can be modified exten-

[1] Supported by NIH grants HL39308 and HL49264.

sively by either supplementing the diet of the tumor-bearing mouse with plant oils, or adding specific PUFA to the medium of cell cultures [1, 2]. Fatty acid compositional changes were observed in membrane fractions isolated from the tumor cell homogenates. In the plasma membrane, the fatty acid changes were not accompanied by any compensatory changes in cholesterol content or phospholipid head group composition [3]. What occurred, then, is a substitution of fatty acyl chains without any major structural reorganization of the membrane lipid bilayer.

Similar fatty acid compositional modifications have been produced in two different human tumor lines in culture, the Y79 retinoblastoma [4] and U937 monocytic leukemia [5]. Both of these tumors grow as a cell suspension in culture. When docosahexaenoic acid (DHA, 22:6) was added to the medium, the 22:6 content of the U937 cells increased from 2.2% in control cultures supplemented with oleic acid to 27.7%, and the total PUFA content increased from 24.9 to 40.4% [5]. Likewise, in the Y79 cells, the 22:6 content of the phospholipids increased from 4.1% in the oleic acid-treated cells to 12.7% in the cultures supplemented with DHA.

Functional Effects of PUFA Enrichment

We initially explored membrane-related functions because substantial fatty acid compositional changes occurred in the tumor cell membranes. PUFA supplementation of Ehrlich ascites carcinoma and L1210 leukemia cells was associated with small increases in plasma membrane fluidity [6, 7]. Some carrier-mediated membrane transport systems were affected [4, 7, 8], suggesting that the fluidity changes occurred in lipid domains that interact with certain membrane transporters and thereby influenced their properties. Differences also were observed in the susceptibility of hepatoma cells to complement-mediated cytolysis and attack by natural killer cells [9, 10].

L1210 cells enriched in DHA were found to be more responsive to treatment with doxorubicin or hyperthermia [1, 2], two forms of therapy that generate free radicals. Therefore, we decided to explore the possibility that DHA enrichment acts by increasing the susceptibility of cancer cells to lipid peroxidation. Using a spin trap, we demonstrated an increase in lipid radical adduct formation when U937 cells enriched with DHA were exposed to Fe^{2+}-mediated oxidant stress [5]. This is illustrated in figure 1. A similar increase was not observed when the U937 cells were enriched with oleic acid. The electron paramagnetic resonance (EPR) spin adduct spectrum produced by the cells supplemented with DHA is consistent with the generation of a lipid radical formed during PUFA oxidation.

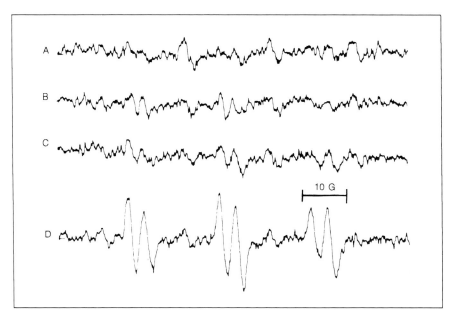

Fig. 1. EPR spectra of the lipid radical adduct produced by U937 human monocytic leukemia cells. The spin trap was α-(4-pyridyl-1-oxide)-N-tert-butylnitrone (POBN). Incubations were done at room temperature in 0.5 ml air-saturated phosphate-buffered saline solutions containing 10 mM POBN and 4×10^6 cells. EPR scans were initiated after addition of 80 μM $FeSO_4$. The spectra shown are: (A) DHA-enriched cells without added $FeSO_4$; (B) cells not enriched with any fatty acid supplement; (C) cells enriched in oleic acid; (D) cells enriched in DHA. [From 5 with permission.]

Parinaric Acid

As an extension of this approach, we tested the ability of *cis*-parinaric acid, a PUFA containing four conjugated double bonds (9,11,13,15–18:4), to sensitize the U937 cells to environmental stress [11]. As shown in figure 2, parinaric acid was cytotoxic in 24-hour incubations at concentrations of 1–4 μM. By contrast, 18-carbon PUFA that do not contain conjugated double bonds, 9,12,15–18:3 and 6,9,12,15–18:4, did not begin to produce cytotoxicity under these conditions until the concentration reached 40–50 μM. Very low concentrations of parinaric acid also were cytotoxic to human Y79 retinoblastoma and HL-60 leukemia cells [11]. Addition of butylated hydroxytoluene to the U937 cultures prevented the cytotoxic action of parinaric acid [11]. This suggests that parinaric acid may exert its effect by sensitizing the tumor cells to ordinary levels of oxidant stress.

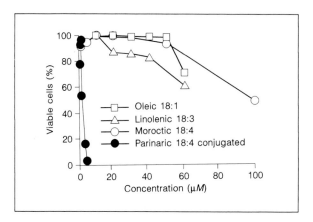

Fig. 2. Cytotoxicity of fatty acids for U937 human monocytic leukemia cells. The cells were incubated with the fatty acids in RPMI-1640 medium containing 5% fetal bovine serum for 24 h. All of the incubation flasks were shielded from light. Cell viability was assessed by trypan blue exclusion. Each point is the mean of three values obtained from separate cultures. [From 11, with permission.]

Therapeutic Implications

Two new approaches to cancer therapy are suggested by these results. One involves the use of DHA supplements to augment the cytotoxicity of existing therapeutic modalities. Based on the data obtained with the human U937 cells, it seems unlikely that DHA, by itself, will produce enough growth inhibition to be therapeutically effective. A more promising strategy would be to load the tumor with DHA in order to enhance the free radical response produced by cytotoxic agents like doxorubicin or ionizing radiation. A second approach is to develop conjugated fatty acids as a new class of chemotherapeutic agent, using parinaric acid as a prototype. Because it is effective at such low concentrations, parinaric acid may be useful as a primary form of therapy.

A critical issue that must be addressed is whether these fatty acid approaches also will sensitize normal tissues, causing an unacceptable increase in generalized toxicity. In addition, it is not known whether either of these approaches are effective in vivo. These questions must be fully explored before further consideration is given to the use of PUFA in cancer therapy.

References

1   Spector AA, Burns CP: Biological and therapeutic potential of membrane lipid modification in tumors. Cancer Res 1987;47:4529–4537.
2   Burns CP, Spector AA: Membrane fatty acid modification in tumor cells: A potential therapeutic adjunct. Lipids 1987;22:178–184.
3   Awad AB, Spector AA: Modification of the fatty acid composition of Ehrlich ascites tumor cell plasma membranes. Biochim Biophys Acta 1976;426:723–731.
4   Yorek MA, Strom DK, Spector AA: Effect of membrane polyunsaturation on carrier-mediated transport in cultured retinoblastoma cells: Alterations in taurine uptake. J Neurochem 1984;42: 254–261.
5   North JA, Spector AA, Buettner GR: Detection of lipid radicals by electron paramagnetic resonance spin trapping using intact cells enriched with polyunsaturated fatty acid. J Biol Chem 1992;267:5743–5746.
6   King ME, Spector AA: Effect of specific fatty acyl enrichments on membrane physical properties detected with a spin label probe. J Biol Chem 1978;253:6493–6501.
7   Burns CP, Luttenegger DG, Dudley DT, Buettner GR, Spector AA: Effect of modification of plasma membrane fatty acid composition on fluidity and methotrexate transport in L1210 murine leukemia cells. Cancer Res 1979;39:1726–1732.
8   Kaduce TL, Awad AB, Fontenelle LJ, Spector AA: Effect of fatty acid saturation on α-aminoisobutyric acid transport in Ehrlich ascites cells. J Biol Chem 1977;252:6624–6630.
9   Yoo T-J, Chiu HC, Spector AA, Whiteaker RS, Denning GM, Lee NF: Effect of fatty acid modifications of cultured hepatoma cells on susceptibility to complement-mediated cytolysis. Cancer Res 1980;40:1084–1090.
10 Yoo T-Y, Kuo C-Y, Spector AA, Denning GM, Floyd R, Whiteaker S, Kim H, Abbas M, Budd TW: Effect of fatty acid modification of cultured hepatoma cells on susceptibility to natural killer cells. Cancer Res 1982;42:3596–3600.
11 Cornelius AS, Yerram NR, Kratz DA, Spector AA: Cytotoxic effect of cis-parinaric acid in cultured malignant cells. Cancer Res 1991;51:6025–6030.

Dr. Arthur A. Spector, Department of Biochemistry, 4-403 BSB. University of Iowa, Iowa City, IA 52242 (USA)

Galli C, Simopoulos AP, Tremoli E (eds): Effects of Fatty Acids and Lipids in
Health and Disease. World Rev Nutr Diet. Basel, Karger, 1994, vol 76, pp 86–88

....

Effects of Eicosapentaenoic Acid on Tumour Growth and Cachexia in Mouse Colon Cancer[1]

M.J. Tisdale, S.A. Beck, E.A. Hudson

CRC Nutritional Biochemistry Research Group, Pharmaceutical Sciences Institute,
Aston University, Birmingham, UK

Growth of some solid tumours is accompanied by the appearance of cachexia, a wasting syndrome characterized by loss of both skeletal muscle and adipose tissue. This syndrome is variably expressed and is dependent on tumour type, but cancers of the pancreas, stomach, lung, liver and kidney display a high incidence of cachexia [1]. Such tumours are resistant to conventional chemotherapy, and surgery remains the optimal modality. We have considered that the process of cachexia may serve as a target for the direction of new types of antitumour agents if the products of catabolism of host tissues are needed by the tumour to satisfy the growth demands. There is considerable background information to suggest that tumours have a specific requirement for certain amino acids in particular glutamine, leucine, methionine and cysteine [2]. In addition, polyunsaturated fatty acids (PUFAs), particularly linoleic (LA) and arachidonic acids (AA), appear to be essential for tumour growth [3] and it is possible that the cachectic process is initiated to satisfy this demand.

As a model system of cachexia we have utilized mice bearing the MAC16 colon adenocarcinoma. This tumour initiates weight loss at tumour volumes >0.3% of the host body weight reaching 30% weight loss when the tumour reaches only 2.5% of host body weight [4]. Moreover, this process occurs without a reduction in either food or water intake and has been attributable to the production by the tumour of a circulatory factor capable of initiating direct catabolism of adipose tissue in vitro [4]. Murine epididymal adipocytes are prepared by collagenase digestion of adipose tissue and mobilization of trigly-

[1] Supported by Programme Grant No. SP 1518 from the Cancer Research Campaign.

cerides is measured by the release of either glycerol or fatty acids. Using such an assay it was found that the PUFA eicosapentaenoic acid (EPA) was an effective inhibitor of lipolysis mediated by an extract of the MAC16 tumour with an IC_{50} value of 40 μM [5].

Stimulation of lipolysis by the tumour factor was associated with an elevation of the intracellular level of cyclic AMP in epididymal adipocytes and this effect was blocked by EPA. Using mouse adipocyte plasma membrane fractions, the MAC16 factor has been shown to directly stimulate adenylate cyclase activity with a time course of increasing activity up to 10 min, after which a decrease in activity was noted in a manner similar to stimulation by polypeptide hormones [6]. This stimulated activity was inhibited by EPA in a dose-dependent manner and most likely arose from the stimulation of an inhibitory guanine nucleotide protein (Gi). The inhibitory effect on tumour-stimulated lipolysis was specific to EPA and was not shown by other related PUFAs of the ω3 or ω6 series.

To determine the usefulness of the in vitro assay for predicting new anticachectic agents in vivo, mice bearing the MAC16 tumour 14 days after transplantation and with an average weight loss of 5% were treated orally by gavage with pure EPA as the free acid and the PUFAs LA and docosahexaenoic acid (DHA) at the same dose level (5 g/kg/day) were used as a control [5]. Weight reversal and re-establishment of the original body weight was seen in the group treated with EPA but not with the other PUFAs. Further studies [7] showed similar weight reversal at dose levels of 1.25–2.5 g/kg/day. Such a dose when translated to cancer patients on a surface area basis would correspond to 6.4–12.8 g/day, assuming an average human surface area of 1.7 m^2. This dose is similar to the average daily intake (7 g) previously reported for Greenland Eskimos [8]. Administration of EPA did not significantly increase the caloric intake of mice, suggesting that the reversal of cachexia arose from direct inhibition of the tumour lipid-mobilizing factor.

Both adipose tissue and skeletal muscle mass were preserved in mice treated with EPA. The effect on skeletal muscle was due to a reduction in the enhanced protein degradation seen in cachectic mice, without an effect on protein synthesis [7].

In addition to the preservation of host body weight, tumour growth was also significantly inhibited by EPA but not by DHA [5, 7]. The antitumour, but not the anticachectic effect of EPA was effectively reversed by co-administration of pure LA [9]. The tumour and plasma level of EPA were not affected by LA. This suggests that tumour growth inhibition may arise as an indirect effect from the inhibition of catabolism of adipose tissue and consequent release of fatty acids. The requirement of the tumour for fatty acids is probably in a regulatory role in growth rather than as an energy supply, since tumour energy metabolism is supplied almost entirely from glucose [10]. Investigation of the kinetics of growth inhibition in mice treated with EPA showed that tumour stasis arose from an increase in the rate of cell loss from 38 to 71% without a change in the potential doubling time [9]. Reversal of the growth inhibitory effect by LA acted to reverse the cell loss factor to 45%.

It therefore seems that some PUFAs are required for tumour growth in vivo to maintain tumour integrity and prevent cell loss. The mechanism of the effect is unknown, but one possibility could be for tumour angiogenesis. Solid tumours often outgrow the blood supply resulting in areas of necrosis and hypoxia. Adipose tissue has long been known to induce neovascularization [11], and recent results have attributed this effect to 12(R)-hydroxy-5,8,14(Z,Z,Z)eicosatrienoic acid formed via metabolism of AA by a cytochrome P-450-dependent pathway [12].

Thus growth of solid tumours may be susceptible to interventions that may not affect growth of normal tissues, thus increasing the selectivity of action. The finding that EPA is effective in inhibiting the growth of a solid tumour that is nonresponsive to conventional cytotoxic chemotherapy may offer an opportunity to treat tumours such as pancreatic adenocarcinomas for which therapy is currently limited.

References

1   Strain AJ: Cancer cachexia in man: A review. Invest Cell Pathol 1979;2:181–193.
2   Tisdale MJ: The future: Nutritional pharmacology; in Calman KC, Fearon KCH (eds): Clinics in Oncology: Nutritional Support for the Cancer Patient. London, Saunders, 1986, vol 5, pp 381–405.
3   Tisdale MJ: Unsaturated fatty acids and cancer; in Unsaturated Fatty Acids: Nutritional and Physiological Significance. London, Chapman & Hall, 1992, pp 120–128.
4   Beck SA, Tisdale MJ: Production of lipolytic and proteolytic factors by a murine tumor-producing cachexia in the host. Cancer Res 1987;47:5919–5923.
5   Tisdale MJ, Beck SA: Inhibition of tumour-induced lipolysis in vitro and cachexia and tumour growth in vivo by eicosapentaenoic acid. Biochem Pharmacol 1991;41:103–107.
6   Adamson CL: Inhibitors of lipolysis in tumour-induced cachexia; PhD thesis, Aston University, Birmingham 1992.
7   Beck SA, Smith KL, Tisdale MJ: Anticachectic and antitumour effect of eicosapentaenoic acid and its effect on protein turnover. Cancer Res 1991;51:6089–6093.
8   Dyerberg J: Linolenate-derived polyunsaturated fatty acids and prevention of atherosclerosis. Nutr Rev 1986;44:125–134.
9   Hudson EA, Beck SA, Tisdale MJ: Kinetics of the inhibition of tumour growth in mice by eicosapentaenoic acid – Reversal by linoleic acid. Biochem Pharmacol 1993;45:2189–2194.
10 Mulligan HD, Tisdale MJ: Metabolic substrate utilization by tumour and host tissues in cancer cachexia. Biochem J 1991;277:321–326.
11 Silverman KJ, Lund DP, Zetter BR, et al: Angiogenic activity of adipose tissue. Biochem Biophys Res Commun 1988;153:347–352.
12 Masferrer JL, Rimarachin JA, Gerritsen ME, et al: 12(R)-hydroxyeicosatrienoic acid, a potent chemotactic and angiogenic factor produced by the cornea. Exp Eye Res 1991;52:417–424.

M.J. Tisdale, CRC Nutritional Biochemistry Research Group, Pharmaceutical Sciences Insitute, Aston University, Birmingham B4 7ET (UK)

Galli C, Simopoulos AP, Tremoli E (eds): Effects of Fatty Acids and Lipids in
Health and Disease. World Rev Nutr Diet. Basel, Karger, 1994, vol 76, pp 89–94

....

ω3 Fatty Acids in the Regulation of Cytokine Synthesis [1]

Stefan Endres, Bhanu Sinha, Tobias Eisenhut

Medizinische Klinik, Klinikum Innenstadt der Ludwig-Maximilians-Universität,
München, Germany

The Cytokines Interleukin-1 and Tumor Necrosis Factor

Cytokines are secreted proteins with central paracrine functions in inflammation and proliferation. They comprise the families of interleukins (IL), tumor necrosis factors (TNF) interferons and the hematopoietic growth factors. Among this large group of proteins, a subgroup, initially identified as products of monocytes, have a similar profile of biological activities. This group comprises the two forms of IL-1, IL-1α and IL-1β, TNF-α and, to a certain extent, IL-6. Among the many biological activities shared by these cytokines, their actions on cells of the vascular endothelium appear to be of particular importance in the pathogenesis of atherosclerosis. Among them, IL-1 potentiates procoagulant activity, increases the production of plasminogen activator inhibitor and endothelin, as well as the formation of eicosanoids. Furthermore, it increases leukocyte adhesion by inducing the expression of adhesion molecules and it promotes endothelial protein permeability.

Aside from their pyrogenic effects, IL-1 and TNF influence a wide array of biological functions (fig. 1) [reviewed in 1]. Many of the biological functions of IL-1 are shared by TNF [2]. The study of the IL-1 system has gained additional momentum by the discovery of the IL-1 receptor antagonist. This protein was identified as an endogenous inhibitor of IL-1 action. It shares sequence homologies with IL-1α and IL-1β, binds to the same receptor as these two proteins, but has no intrinsic activity. Aside from its enormous biological

[1] Supported by grant 07ERG03 7 of the Deutsches Bundesministerium für Forschung und Technologie. B.S. is on a scholarship by the Fondacion Federico, Brazil.

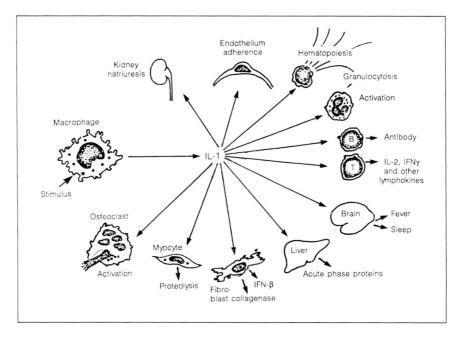

Fig. 1. Biological activities of IL-1 [with courtesy of Dr. Jos W. M. van der Meer, Nijmegen, The Netherlands].

significance, the IL-1 receptor antagonist has proved an excellent experimental tool in demonstrating the necessary role of IL-1 in certain disease processes [reviewed in 3]. Mitigating disease in animal models by administration of exogenous IL-1 receptor antagonist proves participation of IL-1 as a necessary intermediate. This approach has been used in several disease models.

The potent and diverse biological functions of cytokines such as IL-1 and TNF place particular importance on the mechanisms of their regulation and the question of targeted modulation. The synthesis of cytokines is tightly controlled by a host of extracellular factors, plasma membrane constituents, cytosolic and nuclear factors. An example of differential regulation is the pharmacological elevation of intracellular cAMP. This intervention markedly suppresses synthesis of TNF without affecting the synthesis of IL-1 [4].

Modulating the Synthesis of IL-1 and TNF by ω3 Polyunsaturated Fatty Acids

Pharmacologic agents known to reduce the synthesis of IL-1 and TNF synthesis are corticosteroids and cyclosporin A. Since IL-1 and TNF are principal mediators of inflammation, reduced production of these cytokines may contribute to the amelioration of inflammatory symptoms in patients taking ω3 supplementation. Therefore, we have investigated the effects of ω3 fatty acids on the synthesis of the cytokines IL-1 and TNF [5].

Figure 2 illustrates the production of IL-1β during the course of the study. Our findings demonstrate that ω3 fatty acid supplementation reduced the ability of mononuclear cells (MNC) to produce IL-1β upon stimulation with endotoxin. The effect was most pronounced 10 weeks after stopping the supplementation and suggests prolonged incorporation of ω3 fatty acids into a pool of circulating MNC. The capacity of the MNC from these donors to synthesize IL-1β returned to the presupplement level 20 weeks after ending the supplementation. Similar results were observed for IL-1α and TNF.

The findings form a pathophysiologic rationale for therapeutic trials with ω3 fatty acids in certain diseases with documented involvement of inflammatory cytokines such as IL-1 and TNF, like rheumatoid arthritis. Furthermore, a suppression of the magnitude we observed can only be achieved by administration of glucocorticoids or cyclosporin A which have well-known adverse effects, particularly during long-term administration. In an animal transplant model ω3 fatty acids even enhanced cyclosporin A-induced immunosuppression [6].

Other Studies on the Modulation of Cytokine Synthesis by ω3 Fatty Acids

Suppression of IL-1 and TNF synthesis by ω3 fatty acids has been found in other studies in humans. The study design and results are compared in table 1. Of particular interest is the most recent study by Meydani et al. [10], who, for the first time, studied the effect of ω3 fatty acids in dietary intervention (rather than as encapsulated lipid supplement) on cytokine synthesis. After 24 weeks' diet that provided 1.2 g ω3 fatty acids per day, lipopolysaccharide (LPS)-induced synthesis of IL-1β was decreased by 40%.

Recently, studies in animal models have been reported, where dietary ω3 fatty acids did not inhibit, but markedly enhance LPS-stimulated production of TNF [11] or of both IL-1α and TNF [12]. The results of animal studies on the influence of ω3 fatty acids on cytokine synthesis are summarized in table 2. Interestingly, the three studies where enhancement of cytokines synthesis was found were all performed using peritoneal macrophages from mice. In contrast, in the study by Billiar et al. [13] the IL-1 and TNF synthesis of Kupffer cells from rats was suppressed to an extent similar to that observed in humans. Whether this discrepancy within animal models represents a difference in the

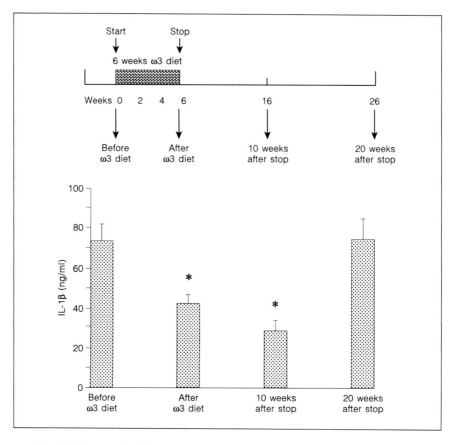

Fig. 2. Influence of ω3 fatty acids on production of IL-1β stimulated with endotoxin (10 ng/ml). Nine healthy volunteers added 18 g of fish oil concentrate (MaxEPA[R]) per day to their normal diet. In vitro production of IL-1β was determined by incubating peripheral blood mononuclear cells for 24 h with different stimuli. At the end of the incubation, the cells were lysed by freeze-thawing to obtain total, that is cell-associated plus secreted cytokine. IL-1β was measured by specific radioimmunoassay. The columns represent the mean values for 9 volunteers. Bars indicate SEM. * Significant difference from the baseline (before ω3 diet) at $p < 0.05$.

cell population studied or a difference in the species is not clear at this time. The mechanism by which ω3 fatty acids affect cytokine synthesis remains unclear. However, studies demonstrating reduced levels of mRNA for platelet-derived growth factor [15] and for IL-1β [16] point to regulation at a pretranslational level.

Table 1. Human studies in vivo examining the effect of ω3 fatty acids on cytokine production

Source	Model	Cells	ω3 fatty acid dose and duration g/day×weeks	Suppression, %	
Endres et al. [5]	Healthy volunteers	MNC	4.6×6	IL-1 61 ↓	TNF 40 ↓
Kremer et al. [7]	Patients with rheumatoid arthritis	MNC	6.7×24	IL-1 41 ↓	
Meydani et al. [8]	Healthy volunteers	MNC	2.4×12	IL-1β 50 ↓	TNF 52 ↓
Mølvig et al. [9]	Healthy volunteers, diabetics	MNC MPh	3.2×7	IL-1β[a] 30 ↓	TNF no Δ
Meydani et al. [10]	Healthy volunteers	MNC	1.2×24	IL-1β 40 ↓	TNF no Δ

[a] Only cell-associated.

Table 2. Animal studies in vivo examining the effect of ω3 fatty acids on cytokine production

Source	Model	Cells	ω3 fatty acid duration, weeks	Change of	
				IL-1	TNF
Billiar et al. [13]	Rats	Kupffer cells	6	↓	↓
Lokesh et al. [14]	Mouse	Peritoneal macrophages	4	↑	↑
Hardardottir et al. [11]	Mouse	Peritoneal macrophages	5	↑	
Blok et al. [12]	Mouse	Peritoneal macrophages	6	↑	↑

References

1   Dinarello CA: The role of interleukin-1 in disease. N Engl J Med 1993;328:106–113.
2   Vassalli P: The pathophysiology of tumor necrosis factors. Annu Rev Immunol 1992;10:411–452.
3   Dinarello CA, Thompson RC: Interleukin-1 receptor antagonist. Immunol Today 1991;12:404–410.
4   Endres S, Fülle HJ, Sinha B, et al: Cyclic nucleotides differentially regulate the synthesis of tumour necrosis factor-α and interleukin-1β by human mononuclear cells. Immunology 1991;72:56–60.
5   Endres S, Ghorbani R, Kelley VE, et al: The effect of dietary supplementation with n-3 polyunsaturated fatty acids on the synthesis of interleukin-1 and tumor necrosis factor by mononuclear cells. N Engl J Med 1989;320:265–271.
6   Kelley VE, Kirkman RL, Bastos M, Barrett LV, Strom TB: Enhancement of immunosuppression by substitution of fish oil for olive oil as a vehicle for cyclosporine. Transplantation 1989;48:98–102.
7   Kremer J, Lawrence D, Jbiz W, et al: Dietary fish oil and olive oil supplementation in patients with rheumatoid arthritis. Clinical and immunological effects. Arthritis Rheum 1990;33:810–820.
8   Meydani SN, Endres S, Woods MM, et al: Oral n-3 fatty acid supplementation suppresses cytokine production and lymphocyte proliferation: Comparison in young and older women. J Nutr 1991;121:547–555.
9   Mølvig J, Pociot F, Worsaae H, et al: Dietary supplementation with ω3 polyunsaturated fatty acids decreases mononuclear cell proliferation and interleukin-1-β content but not monokine secretion in healthy and insulin-dependent diabetic individuals. Scand J Immunol 1991;34:399–410.
10 Meydani SN, Lichtenstein AH, Cornwall S, et al: Immunological effects of National Cholesterol Education Panel (NCEP) step-2 diets with and without fish-derived n-3 fatty acid enrichment. J Clin Invest 1993;92:105–113.
11 Hardardottir I, Kinsella JE: Tumor necrosis factor production by murine resident peritoneal macrophages is enhanced by dietary n-3 polyunsaturated fatty acids. Biochim Biophys Acta 1991;1095:187–195.
12 Blok WL, Vogels MT, Curfs JH, et al: Dietary fish-oil supplementation in experimental gram-negative infection and in cerebral malaria in mice. J Infect Dis 1992;165:898–903.
13 Billiar T, Bankey P, Svingen B, et al: Fatty acid intake and Kupffer cell function: Fish oil alters eicosanoid and monokine production to endotoxin stimulation. Surgery 1988;104:343–349.
14 Lokesh BR, Sayers TJ, Kinsella JE: Interleukin-1 and tumor necrosis factor synthesis by mouse peritoneal macrophages is enhanced by dietary n-3 polyunsaturated fatty acids. Immunol Lett 1990;23:281–285.
15 Kaminski WE, Jendraschek E, Kiefl R, von Schacky C: Dietary ω-3 fatty acids lower levels of PDGF mRNA in human mononuclear cells. Blood 1993;81:1871–1875.
16 Robinson DR: Dietary n-3 fatty acids and inflammatory arthritis (abstract). Abstract book ISSFAL 1993, p 86.

PD Dr. S. Endres, Medizinische Klinik, Klinikum Innenstadt der Universität München, Ziemssenstrasse 1, D-80336 München (Germany)

Galli C, Simopoulos AP, Tremoli E (eds): Effects of Fatty Acids and Lipids in
Health and Disease. World Rev Nutr Diet. Basel, Karger, 1994, vol 76, pp 95–102

....

Alleviation of Autoimmune Disease by ω3 Fatty Acids

Dwight R. Robinson, Li-Lian Xu, Christopher T. Knoell,
Sumio Tateno, Muyi Guo, Robert B. Colvin, Walter Olesiak,
Masaharu Urakaze, Eiji Sugiyama, Philip E. Auron, Edward T.H. Yeh,
K. Frank Austen, Richard I. Sperling

Arthritis Unit of the Medical Services and Department of Pathology,
Massachusetts General Hospital; Department of Rheumatology and Immunology,
Brigham and Women's Hospital, and Departments of Medicine and Pathology,
Harvard Medical School, Boston, Mass., USA

Dietary ω3 fatty acids modify inflammation and benefit inflammatory autoimmune diseases in humans and in experimental animals. Although alterations in the production of inflammatory eicosanoids by ω3 fatty acids, such as the inhibition of leukotriene B_4 (LTB_4) synthesis, could contribute to the anti-inflammatory effects of ω3 fatty acids, other factors may be important, and the mechanisms of the anti-inflammatory effects of ω3 fatty acids remain uncertain [1]. We previously reported that ω3 fatty acids alleviated the severity of autoimmune glomerulonephritis in $[NZB \times NZW]F_1$ (NZB/W), and other inbred strains of mice [2–4]. These autoimmune strains spontaneously develop diseases that resemble the pathologic and serologic changes in the human disease systemic lupus erythematosus (SLE), and the murine disease is considered a good model for human SLE. These and other studies provide the basis for initiating clinical trials of dietary ω3 fatty acids in various human autoimmune diseases, including SLE, rheumatoid arthritis, and psoriasis. As described below, several clinical trials have demonstrated modest, but statistically significant, clinical benefits of dietary supplements of ω3 fatty acids in patients with rheumatoid arthritis. Therefore, we thought that it would be useful to better understand the mechanisms of the anti-inflammatory effects on ω3 fatty acids in order to optimize the use of ω3 fatty acids as therapeutic agents in human autoimmune, and perhaps other, disease states.

Murine Autoimmune Disease and ω3 Fatty Acids

Early studies from our laboratory demonstrated that dietary ω3 fatty acids significantly reduced the severity of autoimmune glomerulonephritis by over 80% compared to controls fed with beef tallow (BT) as a lipid source, when the dietary supplement is administered from the time of weaning up to age 19 months [3]. In order to document the active components of the fish oil (FO) triglyceride that were responsible for the beneficial effects, we carried out a series of experiments that compared BT and FO with the two major ω3 fatty acids in marine lipids, eicosapentaenoic acid (EPA) and docosahexaenoic acid (DHA) separately, and also combined in a single diet [5]. The purified ω3 fatty acids (90%) were in the form of ethyl esters. Each experimental diet consisted of a balanced fat-free powder to which one of the lipid preparations was added. All diets contained 2 wt% safflower oil to avoid essential ω6 fatty acid deficiency, and 10 wt% of either BT or one of the ω3 fatty acid preparations. In the diets containing <10% of either the FO or the ω3 fatty acid ethyl esters, the lipid content was made up to 10% with BT. The experimental diets were initiated at age 22 weeks, prior to the development of significant glomerulonephritis, and were continued for an additional 13 weeks, at which time the BT controls had developed moderately severe disease, as measured by kidney histology, and the mice were sacrificed.

The changes in glomerular pathology are summarized in table 1, utilizing a standard histologic scoring system. We conclude from these results that purified ω3 fatty acids as well as FO triglycerides alleviate the severity of murine autoimmune glomerulonephritis. However, since the 10% FO contains approximately 33% EPA and 12% DHA, the 10% FO diet contains a similar quantity of EPA as does the 3% EPA diet, and neither the latter nor the 6% EPA diets or the 3% DHA diets have any significant protective effects. Thus, neither the EPA nor the DHA in the FO alone can account for the protective effect of FO. Comparison of the EPA and the DHA diets demonstrates that DHA confers a greater degree of benefit on this renal disease than does EPA, when the two are given in the same quantities. Somewhat surprisingly, the diet combining the two ω3 fatty acid ethyl esters suppresses the degree of glomerular capillary thickening. This observation was confirmed in an additional experiment, in which a similar combined ω3 fatty acid ethyl ester diet significantly reduced the severity of several aspects of the renal pathology (data not shown). These results indicate that these two ω3 fatty acids have synergistic beneficial effects on autoimmune renal disease. Thus the combination of long chain ω3 fatty acids in marine lipids may be superior to purified single ω3 fatty acids, but the optimal proportion of each of the fatty acids is unknown at present.

Table 1. Effects of ω3 fatty acids on renal pathology in NZB/W mice [from 5, with permission]

Dietary group		Mice n	Glomerular cellularity[a]	Glomerular capillary thickening[a]	Percent mice with normal glomeruli[b]
BT		35	2.1 ± 0.2	1.8 ± 0.2	17
FO	10%	44	1.0 ± 0.2[c]	0.5 ± 0.1[c]	70[e]
FO	5%	13	2.2 ± 0.2[d]	1.5 ± 0.3[d]	23[d]
EPA-E	3%	13	2.7 ± 0.2[d]	2.0 ± 0.3[d]	8[d]
	6%	12	2.1 ± 0.2[e]	1.4 ± 0.3[d]	8[d]
	10%	15	0.7 ± 0.3[c]	0.3 ± 0.2[c]	73[f]
DHA-E	3%	14	1.7 ± 0.3	1.1 ± 0.3[e]	36
	6%	14	0.7 ± 0.2[c]	0.1 ± 0.1[c]	93[c, e]
	10%	14	0.5 ± 0.2[c]	0.1 ± 0.1[c]	93[c, e]
EPA-E DHA-E	2.8% + 0.8%	11	2.0 ± 0.3[e]	0.7 ± 0.2[f]	27

[a] Severity of glomerular pathology measured scores ranging from 0 (normal) to 3 (most severe). Mean scores ± SEM are listed. In the experiments reported in this table, glomerular mesangial material, sclerosis, and tubulointerstitial changes were insignificant in all groups.
[b] Glomeruli were considered normal glomeruli when the mean cellularity scores were either 0 or 1, and capillary thickening scores were 0, for each individual mouse kidney section. The number of mice with normal glomeruli within a dietary group was divided by the total number of mice in that group (×100) to give the percent of mice with normal glomeruli.
[c] $p < 0.001$ vs. BT.
[d] $p < 0.001$ vs. FO 10%; n.s. vs. BT.
[e] $p < 0.05$ vs. FO 10%; n.s. vs. BT.
[f] $p < 0.02$ vs. BT.

Comparison of FO and ω3 Fatty Acid Ethyl Esters on Phospholipid Fatty Acid Composition

We have compared the tissue phospholipid fatty acid composition from these mice, on all of the major phospholipid classes and subclasses [6]. We chose to analyze spleen phospholipids since lymphocytes and monocytes are abundant in spleen, and these cells are presumed to be important cells in the pathogenesis of autoimmune disease. Some representative data is shown in table 2, for 1-O-alk-1'-enyl-2-sn-acyl-glycero-3-phosphoethanolamine. In spite of the large range in the total polyunsaturated fatty acid (PUFA) contents of the diets, the total PUFA, total monoene, and total saturated fatty acid contents of this and other phospholipid classes remain relatively constant. However, the

Table 2. Fatty acid composition of 1-O-alk-1′-enyl-2-sn-acyl-glycero-3-phosphoethanol-amines from NZB/W mice fed either BT or 10% marine lipid diets; means ± SEM of two determinations [from 6, with permission]

FAME	BT	FO 10%	EPA-E 10%	DHA-E 10%
14:0	0.2 ± 0.2			
16:0	5.3 ± 2.3	6.2 ± 1.2	4.4 ± 0.7	2.5 ± 0.3
18:0	1.7 ± 0.3	3.5 ± 0.4	2.1 ± 0.2	3.3 ± 0.4
16:1ω9	0.1 ± 0.1		1.0 ± 0.4	
16:1ω7	0.1 ± 0.1	2.5 ± 0.4	1.2 ± 0.5	1.0 ± 0.1
18:1ω9	2.6 ± 0.1	3.1 ± 0.5	2.2 ± 0.0	2.5 ± 0.2
18:1ω7	0.7 ± 0.0	0.9 ± 0.1		
18:2ω6	1.2 ± 0.1	1.9 ± 0.2	1.2 ± 0.0	2.2 ± 0.0
20:3ω6				0.5 ± 0.0
20:4ω6	52.2 ± 1.0	18.1 ± 0.7	24.6 ± 0.3	10.5 ± 0.4
22:4ω6	18.7 ± 0.7	1.3 ± 0.0	1.9 ± 0.01	0.6 ± 0.0
22:5ω6	6.3 ± 0.0			0.7 ± 0.0
20:5ω3	0.2 ± 0.0	18.7 ± 0.5	21.7 ± 0.4	12.1 ± 0.1
22:5ω3	1.1 ± 0.0	19.0 ± 0.8	33.5 ± 1.0	7.1 ± 0.1
22:6ω3	9.7 ± 0.1	25.0 ± 2.2	6.3 ± 0.4	57.1 ± 1.1
Saturated	7.2	9.6	6.5	5.8
Monoene	3.5	6.5	4.4	3.5
ω6 PUFA	78.4	21.3	27.7	14.4
ω3 PUFA	10.9	62.7	61.5	76.3
Total PUFA	89.3	84.0	89.2	90.7

composition of the PUFA varies greatly with the different diets. The large quantities of 22 carbon ω6 fatty acids in the BT group are nearly eliminated by each of the three ω3 diets. In addition, large, but less complete, reductions in arachidonic acid (AA) are seen with the ω3 diets, and in contrast, small increases in levels of linoleate are seen with the ω3 diets.

The ω3 diets all result in large increases in long chain ω3 fatty acids. Although extensive elongation of EPA has apparently occurred in mice ingesting the EPA ester diet, no elevation in the levels of DHA is observed. No retroconversion of EPA to shorter chain ω3 fatty acids is observed either. In contrast, extensive retroconversion of DHA to both docosapentaenoic acid (DPA, 22:5 ω3) and EPA is noted, based on the large quantities of DPA and EPA in phospholipids from mice fed the DHA ethyl ester diets. Changes in tissue ω3 fatty acid contents with a diet containing both EPA and DHA ethyl esters are similar to changes seen with FO diets. Thus the synergistic effects of diets combining both EPA and DHA cannot be accounted for by synergy in elevating tissue ω3 fatty acid contents.

The Mechanism of Cytokine Suppression by ω3 Fatty Acids

Dietary supplements of ω3 fatty acids in human volunteers suppress the levels of interleukin-1 (IL-1) and tumor necrosis factor-α (TNF-α) produced by lipopolysaccharide (LPS)-stimulated peripheral blood mononuclear cells ex vivo [see Endres, this volume]. In order to extend these observations and investigate the mechanism of suppression of cytokine formation by ω3 fatty acids, we have studied the mechanisms of the effects of dietary FO on stimulated splenic mononuclear cells from BALB/c mice, which were fed experimental diets containing either 15 wt% BT or FO, the latter containing 55% ω3 fatty acids. We found that after ingesting experimental diets from 6 to 12 weeks, cells stimulated with either LPS or phorbol myristate acetate (PMA) contained lower levels of proIL-1β mRNA as measured by Northern analyses. The stability of the proIL-1β mRNA from either BT- and FO-fed mice were similar, after treating stimulated cells with actinomycin D (t 1/2 approx. 2–3 h), and nuclear run on experiments demonstrated reduced transcription rates for cells from FO-fed mice. We conclude from these experiments that dietary ω3 fatty acids reduce IL-1 gene transcription. Since IL-1 and other cytokines are postulated to contribute to the pathogenesis of autoimmune diseases including rheumatoid arthritis, as well as other pathologic states such as atherosclerosis, the ability of ω3 fatty acids to suppress cytokine gene transcription may contribute to their therapeutic effects.

Dietary ω3 Fatty Acids Inhibit Chemotaxis and Second Messenger Formation

Several studies have demonstrated that dietary ω3 fatty acids inhibit the chemotactic responsiveness of human neutrophils, as well as the production of LTB$_4$ and other 5-lipoxygenase products by ionophore-stimulated neutrophils ex vivo [for review, see 7]. In a recent study of neutrophils from human volunteers who ingested approximately 16 g of ω3 fatty acids, as their ethyl esters, the chemotactic response of neutrophils to LTB$_4$ decreased by over 90%, after 10 weeks of dietary ω3 fatty acids [8]. At the same time, inositol phosphate levels after stimulation with LTB$_4$ were reduced, with 90% reduction in the level of the important second messenger, IP$_3$, compared to levels in stimulated neutrophils prior to the dietary ω3 supplement. Thus, ω3 fatty acids can suppress signal transduction at the level of the PI-specific phospholipase C in humans [8].

Table 3. Clinical trials of ω6 fatty acids in patients with rheumatoid arthritis [modified from 10]

	Kremer	Kremer	Cleland et al.	Kremer
Number of patients	37	33	44	49
Study design	Prospective, double-blinded, placebo-controlled	Prospective, double-blinded crossover	Prospective, double-blinded placebo-controlled	Prospective, double-blinded, placebo controlled with different doses of FO
Type of FO	Triglyceride	Triglyceride	Triglyceride	Ethyl ester
Duration of FO ingestion, weeks	12	14	12	24
Placebo (control)	Paraffin wax	Olive oil	Olive oil	Olive oil
Results	Significant improvement in tender joints and morning stiffness	Significant improvement in tender joints, interval to fatigue onset	Significant improvement in tender joints and grip strength	Multiple significant clinical outcomes, most common in high-dose FO group

	Van Der Tempel et al.	Tulleken et al. [11]	Kjeldsen-Kragh et al. [12]
Number of patients	16	27	27
Study design	Prospective, double-blinded, placebo-controlled	Prospective, double-blinded, controlled	Prospective, double-blinded, placebo-controlled
Type of FO	Triglyceride	Ethyl ester	Triglyceride
Duration of FO ingestion, weeks	12	12	16
Placebo (control)	Coconut oil	Coconut oil	Corn oil
Results	Significant improvement in swollen joints and morning stiffness	Significant improvement in joint pain and joint swelling indices	Significant improvement in morning stiffness and global assessments by patients and physicians

Clinical Trials of ω3 Fatty Acids in Rheumatic Diseases

Following the demonstration that dietary ω3 fatty acids alleviate autoimmune disease and other inflammatory diseases in experimental animals, several clinical trials of dietary supplements of ω3 fatty acids have been carried out in human inflammatory diseases. A study of a small number of patients with SLE found statistically significant improvement in disease activity in patients taking a fish oil triglyceride compared to those taking olive oil, both at 20 g daily for 12 weeks [9]. The crossover design of this study was not ideal, as noted below for rheumatoid arthritis trials, but delayed effects of FO could have masked some of the apparent benefits. This encouraging pilot study supports the need for more definitive, larger scale trials.

We are aware of reports of seven double-blind, placebo-controlled trials of patients with rheumatoid arthritis (table 3). All of these trials have found statistically significant, though modest, beneficial effects on certain parameters that are employed in rheumatoid arthritis trials to reflect clinical benefits. The maximal dose of the sum of EPA and DHA is between 5 and 6 g daily in the previous studies. One study found that patients receiving 5.4 g/day/70 kg body weight of these two fatty acids improved significantly, whereas patients receiving one half of that dose failed to show any benefit.

Clinical improvement was not generally noted until after at least 12 weeks of treatment, with optimal effects observed after 18–24 weeks. In addition, sustained improvement was observed for several weeks after discontinuing the FO supplements, making the convenient cross-over design of such studies inadvisable. As in other human experiments, toxicity of dietary supplements of ω3 fatty acids have been minimal. The most frequent side effects have been abdominal discomfort and occasional loose stools, and unpleasant eructation. Although slight prolongation of bleeding time tests occurred, abnormal bleeding has not been a clinical problem. The best choice of placebo material has not been resolved, and authors of two of the reported studies have suggested that the olive oil placebo may have actually had beneficial effects, and if so, this would tend to minimize the observable beneficial effect of the ω3 fatty acid preparation in some studies.

In spite of these apparent beneficial effects of dietary ω3 fatty acids in patients with rheumatoid arthritis, this intervention has not found general acceptance as a treatment modality for this disease. The problem lies in that the beneficial effects, while statistically significant, were modest enough to be of little practical use to patients. It is possible that greater benefit would be seen with larger doses of ω3 fatty acids. Another possible means of enhancing the beneficial effects of ω3 fatty acids would be to prolong the duration of therapy beyond 24 weeks, since previous studies have shown that the effects require relatively long periods of ω3 fatty acid supplements in order to observe significant effects. Finally, clinical benefits of therapeutic agents are notoriously relatively long periods of ω3 fatty acid supplements in order to observe significant effects. Finally, clinical benefits of therapeutic agents are notoriously difficult to demonstrate in rheumatoid arthritis because of several factors, including the variable course of the disease, the crude nature of the clinical parameters available to estimate the state of the disease, and the susceptibility of the disease to improvement from placebo effects. It is apparent that large-scale trials of long duration will be required in order to determine the role of ω3 fatty acids in therapy of rheumatoid arthritis, and probably also for other rheumatic diseases. Since adequate therapy is lacking for many patients with these important diseases, and since ω3 fatty acids are apparently free of serious side effects, unlike most existing therapies, such large-scale clinical trials are needed.

References

1   Robinson DR: Alleviation of autoimmune disease by dietary lipids containing omega-3 fatty acids. Rheum Dis Clin North Am 1991;17:213.
2   Prickett JD, Robinson DR, Steinberg AD: Dietary enrichment with the polyunsaturated fatty acid eicosapentaenoic acid prevents proteinuria and prolongs survival in NZB × NZW/F$_1$ mice. J Clin Invest 1981;68:556–559.
3   Prickett JD, Robinson DR, Steinberg AD: Effects of dietary enrichment with eicosapentaenoic acid upon autoimmune nephritis in female NZB × NZW/F$_1$ mice. Arthritis Rheum 1983;26:133–139.
4   Robinson DR, Prickett JD, Makoul GT, Steinberg AD, Colvin RB: Dietary fish oil reduces progression of established renal disease in (NZB × NZW)F$_1$ mice and delays renal disease in BXSB and MRL/1 strains. Arthritis Rheum 1986;29:539.
5   Robinson DR, Xu Li-L, Tateno S, Guo M, Colvin RB: Suppression of autoimmune disease by dietary n-3 fatty acid. J Lipid Res 1993;34:1435.
6   Robinson DR, Xu Li-L, Knoell CT, Tateno S, Olesiak W: Modification of spleen phospholipid fatty acid composition by dietary fish oil and by n-3 fatty acid ethyl esters. J Lipid Res 1993;34:1423.
7   Sperling RI: Dietary omega-3 fatty acids: Effects on lipid mediators of inflammation and rheumatoid arthritis. Rheum Dis Clin North Am 1991;17:373.
8   Sperling RI, Benincaso AI, Knoell CT, Larkin JK, Austen KF, Robinson DR: Dietary ω-3 polyunsaturated fatty acids inhibit phosphoinositide formation and chemotaxis in neutrophils. J Clin Invest 1993;91:651–660.
9   Walton AJE, Snaith ML, Locniskar M, Cumberland AG, Morrow WJW, Isenberg DA: Dietary fish oil and the severity of symptoms in patients with systemic lupus erythematosus. Ann Rheum Dis 1991;50:463–466.
10 Kremer JM: Clinical studies of omega-3 fatty acid supplementation in patients who have rheumatoid arthritis. Rheum Dis Clin North Am 1991;17:391.
11 Tulleken JE, Limburg PC, Muskiet FAJ, van Rijswijk MH: Vitamin E status during dietary fish oil supplementation in rheumatoid arthritis. Arthritis Rheum 1990;33:1416–1419.
12 Kjeldsen-Kragh J, Lund JA, Riise T, Finnanger B, Haaland K, Finstad R, Mikkelsen K, Forre O: Dietary omega-3 fatty acid supplementation and naproxen treatment in patients with rheumatoid arthritis. J Rheumatol 1992;19:1531–1536.

Dwight R. Robinson, MD, Arthritis Unit, Bulfinch 165, Massachusetts General Hospital, Boston, MA 02114 (USA)

Galli C, Simopoulos AP, Tremoli E (eds): Effects of Fatty Acids and Lipids in
Health and Disease. World Rev Nutr Diet. Basel, Karger, 1994, vol 76, pp 103–104

....

Summary Statement: Essential Fatty Acids, Pregnancy and Pregnancy Complications

Gerard Hornstra

The session was co-chaired by *G. Hornstra* and *S.W. Walsh,* and presentations were made by Drs. *Walsh, M.D.M. Al, G. Vilbergsson, M.M.H.P. Foreman-van Drongelen, J.D. Sorensen,* and *M.C. Craig-Schmidt.*

Although it is generally accepted that essential fatty acids (EFA) play a crucial role in fetal development and pregnancy outcome, surprisingly little information is available on the EFA status of mother and fetus during normal and complicated pregnancy. Data on this topic are now gradually becoming available, and in this session new data were presented, demonstrating that pregnancy complications like intrauterine growth retardation, pregnancy-induced hypertension, and preeclampsia are associated with distinct alterations in both the maternal as well as the neonatal EFA status. In addition, increased lipid peroxidation in preeclampsia, that seems tightly coupled to placental cyclooxygenase activity and thromboxane production, was reported. Stimulation of cyclooxygenase results in increased production of both thromboxane and lipid peroxides, whereas inhibition of cyclooxygenase with low-dose aspirin results in decreased production of both. The human placenta secretes lipid peroxides, and so is a likely source of lipid peroxides in the maternal circulation. In the placental circulation, lipid peroxides produce vasoconstriction by stimulating thromboxane. The question whether these alterations are merely an effect of these pregnancy complications, or possibly contribute to their cause, requires prospective longitudinal studies which are difficult to perform. Nonetheless, this approach has already proven to be successful. An increased content of 22:6ω3, DHA, was observed in plasma phospholipids of women suffering from pregnancy-induced hypertension (PIH), which is likely to be consequential since it developed *after* the onset of hypertension. The same longitudinal study, which is still being continued, also indicates that alterations in the maternal AA status may be causal to PIH, since these alterations are observed at 16 weeks of pregnancy.

The neonatal EFA status at birth has a rather long-lasting effect on the postnatal EFA status. Especially in preterm infants, the EFA status at birth may be even of greater importance than the EFA content of formula. Although longer term studies need to be performed to fully explore this phenomenon, data point to the importance of optimizing the EFA content of the maternal diet during gestation, in order to improve the EFA status of the neonate.

It was demonstrated that the daily supplementation of pregnant women with 2.7 g ω3 PUFA from fish oil, starting at the 30th week of pregnancy, results in significant decreases in thromboxane A_2 (measured as TxB_2) and increases in the urinary excretion of the major metabolites of prostacyclins I_2 and I_3. When related to the changes in maternal and neonatal fatty acid profiles, an inverse correlation was found between the ω3 status and the formation of TxA_2. In addition, a positive relationship was observed between the ω3 status and the urinary excretion of the major metabolite of prostaglandin I_3. These observations may provide an explanation for the slight (4 days) but significant prolongation in the duration of pregnancy, observed in the same study and recently described by others.

The studies reported during this session strongly suggest that EFA supplementation during pregnancy is likely to affect pregnancy outcome. It is to be hoped that more prospective, longitudinal studies will be performed to better document the possible importance of the maternal EFA status during gestation for an optimal EFA status of the neonate and its consequences in later life.

Acknowledgement

The financial support of Nutritia BV, Zoetermeer, The Netherlands, is gratefully acknowledged.

Galli C, Simopoulos AP, Tremoli E (eds): Effects of Fatty Acids and Lipids in
Health and Disease. World Rev Nutr Diet. Basel, Karger, 1994, vol 76, pp 105–109

....

Essential Fatty Acid Status Is Altered in Pregnancies Complicated by Intrauterine Growth Retardation

*G. Vilbergsson, M. Wennergren, G. Samsioe, P. Percy, A. Percy,
J.-E. Månsson, L. Svennerholm*

Departments of Obstetrics and Gynecology, and Psychiatry and Neurochemistry,
University of Göteborg, Sweden

Introduction

Intrauterine growth retardation (IUGR) is still one of the major causes of
fetal morbidity and mortality in the perinatal period. The growth-retarded
newborn runs a substantial risk of later developmental disorders and
handicaps. Neurodevelopmental performance is often affected, as is future
somatic growth.

In IUGR the fetal blood flow is diminished as is the blood flow over the
placenta. The growth-retarded fetus is often hypoxic and hypoglycemic already
before delivery and the truly IUGR fetuses are often suffering from malnutrition.

IUGR is often most pronounced during the third trimester of the pregnancy when there is an increased demand for a steady flow of nutrients to the
developing fetus. Essential fatty acids (EFA) and their long chain polyunsaturated derivatives are also an important part of structural lipids like cell
and organelle membranes. During fetal life the brain accumulates increasing
amounts of these fatty acids which are mandatory for normal function. They
also play a role in maintaining the integrity of the vascular system, on which the
brain depends [1, 2].

Analysis of the relative fatty acid composition of umbilical vein and
arterial tissue in normal pregnancies showed clearly lower values of EFA in the
efferent vessels [3, 4]. In preeclampsia, a situation often complicated by IUGR,
a derangement of EFA has been described [5–7] and this derangement has been
said to be among the causes of decreased fetal blood flow via unbalanced

prostaglandin production [5, 8, 9]. In two of the studies on preeclampsia it is not mentioned whether the pregnancies studied were complicated by fetal growth retardation [5, 6]. The EFA status in IUGR pregnancies does not seem to have been analyzed in detail.

Material and Methods

In order to delineate the EFA status in IUGR the fatty acid composition of phosphatidylcholine (PC) in maternal plasma and in cord blood of 13 cases of IUGR pregnancies was compared to that of 20 controls. Blood drawn from the mothers as well as from the umbilical cord was put in EDTA tubes then centrifuged and the serum was recovered and frozen for later analysis. The fatty acid composition in total phosphoglycerides and in the major phosphoglycerides, PC and phosphatidylethanolamine (PE) of 15 placentas from another group of IUGR pregnancies was also compared with that of 7 control placentas. The placenta was immediately put on ice and transported to the laboratory within 2–3 h. The placenta was washed and pressed until the supernate was free from visible blood. Extra fluid was removed from the tissue. Analyses of dry weight and the fatty acid composition of phospholipids PC and PE were performed. The separation and isolation of the phospholipids PC and PE was carried out as described by Olegård and Svennerholm [10]. After hydrolyzation and subsequent methylation the fatty acid methyl esters were analyzed on chromatographs. The relative fatty acid content was determined by calculation of the peak area on the chromatogram. The relative fatty acid compositions are given as mol%. For a comprehensive description of the methods used, see Percy et al. [11] and Vilbergsson et al. [12].

Student's t test (comparison of two independent variables) and statistical analysis with regression analysis and analysis of variance was carried out using the SAS/STAT computer programme.

Results

The mean birth weight of the IUGR babies was around −3 SD and the gestational age 36 weeks. The controls did not differ from the normal population. In PC in maternal blood there was a reduction in the $\omega 6$ fatty acids 18:2 and 20:4 as well as 22:6ω3 when compared with the controls. In cord blood of the IUGR babies, 20:3ω6, 20:4ω6, and 22:6ω3 acids were lower with a concomitant rise in 18:1ω9. The IUGR infants had though higher levels of linoleic acid (18:2ω6) at birth than the controls. Despite these elevated linoleic levels there were lower levels of the long chain fatty acid derivatives of the $\omega 6$ family.

The analysis of the placentas revealed no differences regarding dry weight (g/100 g wet weight) or glyceride fatty acid content. In the IUGR placentas the glyceride lipid $\omega 6$ fatty acids including 20:3 and 20:4 were significantly lower. The reduction in $\omega 6$ fatty acids was even more striking in PC with lower values of 20:3ω6 and 20:4ω6 as well as the sum of the ω6 fatty acids. The proportions of 16:0 and 18:0 fatty acids were greater in IUGR placentas. In addition, the PC ω3 fatty acids including 22:6 were significantly less in IUGR. The fatty acid content of PE did not differ from the controls.

Discussion

The fetus is dependent on placental transport of EFA which are incorporated in phospholipids in cell and organelle membranes. The low weight or IUGR preterm infant is especially vulnerable in many ways. There is evidence that these infants require EFA in their immediate postnatal life [1, 2, 13].

EFA deficiency can lead to depletion of the EFA pools in the body, but the various fatty acids can also modulate each other's metabolisms [14]. The extent of this influence is probably different for the different tissues and phospholipid fractions involved [14–16]. Experimental depletion of these acids is associated with functional distortions and high perinatal mortality [13]. A great deal of EFA activity arises from their conversion to the long chain derivatives. The concentration of linoleic acid diminishes and that of the long chain derivatives increases progressively from the maternal liver to placenta, fetal liver and finally the fetal brain. This is what Crawford et al. [1] called biomagnification. The importance of 22:6ω3 for fetal and postnatal development of the central nervous system has been pointed out [13, 17, 18].

The desaturation of EFA is under the complex influence of nutritional and hormonal factors [19] as well as various diseases and condition [20]. Signs of amino acid deficiency have also been described in the growth-retarded fetus [21, 22] and this can diminish the activity of the Δ6-desaturase leading to lower concentration of PUFA even when enough substrate is at hand [19].

One of the suspected factors behind the decreased umbilical blood flow and signs of increased resistance in the umbilical vessels in preeclampsia is a disturbed balance between thromboxane and prostacyclin [9]. Jogee et al. [23] have described reduced prostacyclin synthesis in placenta cells in culture from IUGR pregnancies. Others, however, have questioned this, and Jeremy et al. [24] did not find that placenta produces any significant amount of prostacyclin, although the umbilical vessels were able to produce significant amounts of prostacyclin, more so in the umbilical vein than the artery. Ylikorkala et al. [9] did not find any correlation of maternal levels of thromboxane or prostacyclin to blood flow in the umbilical vein, as measured using Doppler blood flow. Walsh [25] found that placenta produces seven times more thromboxane than prostacyclin. Ongari et al. [26] found evidence for association of EPA status both with IUGR and with fetal vascular prostacyclin production. This indicates that the umbilical cord blood flow is susceptible to vasodilating and antiplatelet aggregatory effects of prostacyclin produced on the fetal side and the prostacyclin/thromboxane balance could be tipped over in an unfavorable direction in IUGR.

There is much that indicates EFA deficiency in IUGR and the changes we have found in the membrane phospholipids of the placenta could affect the export of EFA as well as the transport of other important nutrients to the fetal compartment, or with regard to arachidonic acid ($20:4\omega6$) could alter prostaglandin formation. These findings could explain the hemodynamical changes found in IUGR as well as in these infants' increased vulnerability and poorer performance in neurodevelopmental tests.

References

1   Crawford MA, Hassam AG, Stevens PA: Essential fatty acid requirements in pregnancy and lactation with special reference to brain development; in Holman RT (ed): Progress in Lipid Research, ed 20. New York, Pergamon Press, 1981, pp 19:31–40.
2   Crawford MA: The role of essential fatty acids in neural development: Implications for perinatal nutrition. Am J Clin Nutr 1993;57:703–709.
3   Hornstra G, Van Houwelingen AC, Simonis M, Gerrard JM: Fatty acid composition of umbilical arteries and veins: Possible implication for the fetal EFA status. Lipids 1989;24:511–517.
4   Hornstra G, van der Schouw YT, Bulstra-Ramakers MT, Huisjes HJ: Biochemical EFA status of mothers and their neonates after normal pregnancy. Early Hum Dev 1990;24:239–248.
5   Ogburn PL Jr, Williams PP, Johnson SB, Holman RT: Serum arachidonic acid levels in normal and preeclamptic pregnancies. Am J Obstet Gynecol 1984;148:5–9.
6   Rosing U, Johnson P, Ölund A, Sarnsioe G: The fatty acid composition of serum lecithin after pregnancy complicated by preeclampsia. Arch Gynecol 1984;236:109.
7   Wang YP, Kay HH, Killam AP: Decreased levels of polyunsaturated fatty acids in preeclampsia. Am J Obstet Gynecol 1991;164:812–818.
8   Gude NM, Rica GE, King RG, Boura AL, Brennecke SP: Analysis of the responses of the fetal vessels of human perfused placental lobules to acute infusion of arachidonic acid. Reprod Pert Dev 1990;2:591–596.
9   Ylikorkala O, Jouppila P, Kirkinen P, Viinikka L: Maternal thromboxane, prostacyclin, and umbilical blood flow in humans. Obstet Gynecol 1984;63:677–680.
10 Olegård R, Svennerholm L: Fatty acid composition of plasma and red cell phosphoglycerides in full term infants and their mothers. Acta Paediatr Scand 1970;59:637–641.
11 Percy P, Vilbergsson G, Percy A, Månsson J-E, Wennergren M, Svennerholm L: The fatty acid composition of placenta tissue in intrauterine growth retardation. Biochim Biophys Acta 1991; 1084:173–177.
12 Vilbergsson G, Wennergren M, Samsioe G, Karisson K: Essential fatty acids in pregnancies complicated by intrauterine growth retardation. Int Gynaecol Obstetr 1991;36:277–286.
13 Clandinin MT, Chappel JE, Heim T, Sawyer RR, Chance GW: Fatty acid accretion in the development of human spinal cord. Early Hum Dev 1981;5:1–6.
14 Adam O, Wolfram G, Zöllner N: Effect of α-linolenic acid in the human diet on linoleic acid metabolism and prostaglandin biosynthesis. J Lipid Res 1986;27:421–426.
15 Alling C: Essential fatty acids, malnutrition and brain development; Thesis, Götteborg, 1974.
16 Bruce Å: Changes in the concentration and fatty acid composition of phospholipids in rat skeletal muscle during postnatal development. 1974;90:743–749.
17 Bazan NG: in Wurtman RJ, Wurtman JJ (eds): Nutrition and the Brain. New York, Raven Press, 1990, vol 8, pp 1–24.

18 Lin DS, Connor WE, Anderson GI, Neuringer MJ: Effects of dietary n-3 fatty acids on the phospholipid molecular species of monkey brain. Neurochemistry 1990;55:1200–1207.
19 Brenner RR: Nutritional and hormonal factors influencing desaturation of essential fatty acids, in Holman RT (ed): Progress in Lipid Research, ed 20. New York, Pergamon Press, 1981, pp 41–47.
20 Holman RT, Johnson S: Changes in essential fatty acid profile of serum phospholipids in human disease; in Holman RT (ed): Progress in Lipid Research, ed 20. New York, Pergamon Press, 1981, pp 67–73.
21 Cetin I, Corbetta C, Sereni L, Marconi AM, Bozzetti P, Pardi G, Battaglia F: Umbilical amino acid concentrations in normal and growth-retarded fetuses sample in utero by cordocentesis. Am J Obstet Gynecol 1990;162:253–261.
22 Jansson T, Persson E: Placental transfer of glucose and amino acids in intrauterine growth retardation: Studies with substrate analogs in the awake guinea pig. Pediatr Res 1990;28:203–208.
23 Jogee M, Myatt L, Elder MG: Decreased prostacyclin production by placental cells in culture from pregnancies complicated by fetal growth retardation. Br J Obstet Gynaecol 1983;90:247–250.
24 Jeremy JY, Barradas MA, Craft IL, Mikhailidis DP, Dandona P: Does placenta produce prostacyclin? Placenta 1985;6:45–52.
25 Walsh SY: Preeclampsia: An imbalance in placental prostacyclin and thromboxane production. Am J Obstet Gynecol 1985;152:335–340.
26 Ongari MA, Ritter JM, Orchard MA, Waddel KA, Blair IA, Lewis PJ: Correlation of prostacyclin synthesis by human umbilical artery with status of essential fatty acid. Am J Obstet Gynecol 1984; 149:455–460.

G. Vilbergsson, Departments of Obstetrics and Gynecology, and Psychiatry and Neurochemistry, University of Göteborg, S–400 33 Göteborg (Sweden)

Galli C, Simopoulos AP, Tremoli E (eds): Effects of Fatty Acids and Lipids in
Health and Disease. World Rev Nutr Diet. Basel, Karger, 1994, vol 76, pp 110–113

....

The Relationship between the Essential Fatty Acid Status of Mother and Child and the Occurrence of Pregnancy-Induced Hypertension

Intermediate Results of a Prospective Longitudinal Study

Monique D.M. Al[a], *A.C. v. Houwelingen*[a], *T.H.M. Hasaart*[b], *G. Hornstra*[a]

Departments of [a]Human Biology and [b]Gynaecology and Obstetrics, University of Limburg, Maastricht, The Netherlands

Approximately 10% of all pregnancies are complicated by pregnancy-induced hypertension (PIH). Much of the etiology of PIH is still unknown, but it has been suggested that the eicosanoid system is involved in its pathophysiology. In PIH an imbalance between the eicosanoids thromboxane and prostacyclin exists which favors the actions of thromboxane [1]. The precursors of these eicosanoids are polyunsaturated fatty acids, of which arachidonic acid ($20:4\omega6$, AA) is the main precursor. The ultimate precursor of AA is the essential fatty acid (EFA) linoleic acid ($18:2\omega6$, LA), which cannot be synthesized by man and must, therefore, be derived from the diet.

A previous study has shown decreased maternal levels of LA in plasma phospholipids (PLs) in women with PIH compared to a normal pregnancy (NP) [2]. This was also observed by Wang et al. [3], who studied the fatty acid composition of total lipids from plasma of 9 preeclamptic and 11 normal pregnant women. In contrast, Ogburn et al. [4] did not observe different LA levels in plasma PLs, cholesterol-esters, triglycerides and free fatty acids between preeclamptic and normal pregnant women. However, he did find significantly lower relative amounts of AA in triglycerides and free fatty acids of preeclamptic cord plasma compared to normal cord plasma.

These studies share the disadvantage that they are retrospective case-control studies with a small number of subjects and using material collected in late pregnancy or postpartum. As a consequence, no information is available whether the alteration of the EFA status took place *before* or *after* the development of PIH. Therefore, a prospective, longitudinal study was started to investigate the relationship between the EFA status of mother and child and the occurrence of PIH.

Materials and Methods

Pregnant women, healthy at the start of the study, were asked to give a blood sample before 16 weeks, at 22 weeks and at 32 weeks of gestation. Immediately after delivery, a blood sample from the mother, a blood sample from the umbilical vein and a piece of the umbilical cord were obtained. Blood was collected into tubes containing disodium EDTA as an anticoagulant. Fatty acid compositions (% of total fatty acids) were determined of the PLs from plasma and red blood cells (RBCs) and from the umbilical arterial and venous vessel walls. PIH is defined as an initial diastolic blood pressure below 90 mm Hg, an increase during pregnancy of at least 25 mm Hg or two consecutive measurements of 90 mm Hg or more, 4 h or more apart, or one measurement of 110 mm Hg or more starting after 20 weeks of gestation [5]. The results of every woman with PIH were compared with the results of 3 healthy pregnant women, matched on parity and time and place of birth. Because of the significant correlations observed between several fatty acids and gestational age, in all statistical analyses corrections for gestational age were made.

Results and Discussion

Of the 43 women with PIH collected so far, 36 were nulliparous at entry. Between the two groups the expected differences in gestational age, birthweight and Apgar score (after 1 min) were found, with lower values in the PIH group. In the PIH group, 15 women had proteinuria (>0.3 g/l).

At delivery, but not during gestational weeks 16–32, the relative amounts of docosahexaenoic acid ($22:6\omega3$, DHA) in maternal plasma PLs were significantly higher in PIH as compared to NP. As compared to the situation at 32 weeks, the DHA status at delivery increased in PIH whereas it decreased in NP. The mean relative amount of DHA in umbilical plasma PLs was also significantly higher after a pregnancy with PIH than after NP. No significant differences between the groups were observed for the relative amounts of DHA in maternal and umbilical RBCs and in umbilical arterial and venous vessel wall PLs. Ancheschi et al. [6] also observed similar fatty acid levels in RBC PLs between 6 women with PIH, 4 preeclamptic and 10 normal pregnant women. Because the fatty acid composition in PLs of RBCs and vessel wall is likely to represent a longer term reflection of the EFA status than plasma PLs, it seems that the higher DHA levels in maternal and umbilical PLs at delivery after PIH is a late effect. It is possible that the amount of DHA in plasma PLs rises as a systemic response to stress, possibly associated with PIH. When rats are subjected to repeated administration of high doses of noradrenaline, which is a model for severe stress, the relative amount of DHA of heart muscle PLs increased by approximately 45% [7]. Moreover, catecholamines, which are released as a response to stress, are increased during delivery [8] and increased plasma levels are observed in women with preeclampsia [9, 10]. In contrast to earlier results, the lower content of LA in maternal plasma PLs in PIH did not yet reach significance.

PIH may be the result of a number of different causes with different pathologic characteristics. Therefore, a differential analysis was performed between the more and less severe cases of PIH. Women with an increase of diastolic blood pressure of 40 mm Hg or more (40+) were defined as the more severe cases and women with an increase of less than 40 mm Hg (40−) as the less severe cases. No difference in the maternal LA status was observed between the more and less severe cases. However, the relative amounts of AA in maternal plasma PLs at delivery were significantly lower in the 40+ group compared to their matched controls and significantly higher in the 40− group in comparison to their controls. This contrasting result suggests that indeed different pathological mechanisms may be involved. However, the number of severe cases studied was small and additional data have to be collected before drawing further conclusions.

In summary, these intermediate results have shown that the relative amounts of 22:6ω3 in maternal and umbilical venous plasma PLs at delivery are increased in PIH. It remains to be investigated whether this is due to stress or some other factor. The relative amounts of AA in maternal plasma PLs are decreased in PIH with a diastolic blood pressure increment of 40 mm Hg or more, and increased in PIH associated with less than 40 mm Hg, both compared to a normal pregnancy.

References

1   Walsh SW: Pre-eclampsia: An imbalance in placental prostacyclin and thromboxane production. Am J Obstet Gynecol 1985;152:335–340.
2   Schouw vd YT, Al MDM, Hornstra G, et al: Fatty acid composition of serum lipids of mothers and their babies after normal and hypertensive pregnancies. Prostaglandins Leukot Essent Fatty Acids 1991;44:247–252.
3   Wang Y, Kay HH, Killam AP: Decreased maternal levels of polyunsaturated fatty acids in preeclampsia. Am J Obstet Gynecol 1991;164:812–818.
4   Ogburn PL, Turner SI, Williams PP, et al: Essential fatty acid patterns in preeclampsia. Zentralbl Gynäkol 1986;108:983–989.

5   Davey DA, MacGillivray I: The classification and definition of the hypertensive disorders of pregnancy. Am J Obstet Gynecol 1988;158:892–898.

6   Ancheschi MM, Coata G, Cosmi EV, et al: Erythrocyte membrane composition in pregnancy-induced hypertension: Evidence for an altered lipid profile. Br J Obstet Gynaecol 1992;99:503–507.

7   Emilsson A, Gudbjarnason S: Changes in fatty acyl chain composition of rat heart phospholipids induced by noradrenaline. Biochim Biophys Acta 1981;664:82–88.

8   Hytten F, Chamberlain G (eds): Clinical Physiology in Obstetrics. London, Blackwell Scientific, 1980, p 419.

9   Moodley J, McFadyen ML, Dilray A, et al: Plasma noradrenaline and adrenaline levels in eclampsia. South Afr Med J 1991;80:191–192.

10 Sammour MB, Ammar AR, Tash F, et al: Plasma catecholamines during labour in normal and pre-eclamptic pregnancies; in Bonnar J, MacGillivray I, Symonds M (eds): Pregnancy Hypertension. Proc First Congr ISSHP, 1980, pp 167–173.

M.D.M. Al, MSc, Department of Human Biology, University of Limburg,
PO Box 616, NL–6200 MD Maastricht (The Netherlands)

Galli C, Simopoulos AP, Tremoli E (eds): Effects of Fatty Acids and Lipids in
Health and Disease. World Rev Nutr Diet. Basel, Karger, 1994, vol 76, pp 114–118

....

The Role of Fatty Acid Peroxidation and Antioxidant Status in Normal Pregnancy and in Pregnancy Complicated by Preeclampsia [1]

Scott W. Walsh

Departments of Obstetrics and Gynecology, and Physiology,
Medical College of Virginia, Virginia Commonwealth University,
Richmond, Va., USA

This brief review will focus on what is currently known about lipid peroxides in normal pregnancy and pregnancies complicated by *preeclampsia* (i.e., proteinuric hypertension). Antioxidants will also be discussed because they are the compounds that oppose the toxic actions of lipid peroxides and limit the amount of lipid peroxides that are formed.

Normal Pregnancy

Several investigators have reported that maternal levels of lipid peroxides, vitamin E, and antioxidant activity are increased in normal pregnancy as compared with nonpregnancy [1–7]. This indicates that both peroxidation and antioxidation reactions are enhanced during pregnancy. Serum levels of lipid peroxides remain relatively stable throughout gestation, but the levels of vitamin E progressively increase [6]. The progressive increase in vitamin E levels in relationship to the stable levels of lipid peroxides results in a progressive increase in the ratio of vitamin E to lipid peroxides, so there is a gradual favoring of antioxidant activity over peroxidation with advancing gestation in normal pregnancy.

[1] Supported in part by HD 20973 from the National Institute of Child Health and Human Development.

Preeclamptic Pregnancy

As compared to normal pregnancy, several investigators have reported that maternal blood levels of lipid peroxides are significantly elevated in preeclampsia [1–3, 8]. Maternal levels of lipid peroxides are significantly increased in mild preeclampsia and even further increased in severe preeclampsia [8]. Placental tissue levels and production rates of lipid peroxides are also abnormally increased in preeclampsia [9].

In contrast to lipid peroxides, antioxidant activity in maternal blood is significantly decreased in preeclampsia as compared to normal pregnancy [7, 8]. Vitamin E levels, for example, are unaltered in mild preeclampsia, but markedly decreased in severe preeclampsia. Placental tissue levels of vitamin E are also decreased in preeclampsia.

Placental Production and Secretion of Lipid Peroxides

The source of the increased maternal circulating levels of lipid peroxides in preeclampsia is most likely the placenta because the placenta produces and secretes lipid peroxides [9–11], and maternal levels of lipid peroxides decrease after delivery of the placenta [2]. In preeclampsia, placental tissue levels and production rates of lipid peroxides are abnormally increased [9].

The Cyclooxygenase Enzyme

Mechanism for the Coupling of Lipid Peroxide and Thromboxane Production

Preeclampsia is associated with increased production of thromboxane [12], as well as increased production of lipid peroxides. We hypothesized that increased thromboxane may be coupled with increased lipid peroxides by the cyclooxygenase enzyme by the mechanism shown in figure 1 [13]. If this hypothesis is correct, then stimulation of cyclooxygenase should result in increased production of both thromboxane and lipid peroxides, whereas inhibition of cyclooxygenase should result in decreased production of both. The following studies provide proof of this hypothesis.

Stimulation of cyclooxygenase with exogenous peroxide results in increased secretion of both thromboxane and lipid peroxides by the isolated perfused human placental cotyledon [11, 14]. When cyclooxygenase is first inhibited by perfusing the placenta with low-dose aspirin, exogenous peroxide is no longer able to stimulate placental secretion of either thromboxane or lipid peroxides. Placental production, as well as secretion, is inhibited by aspirin.

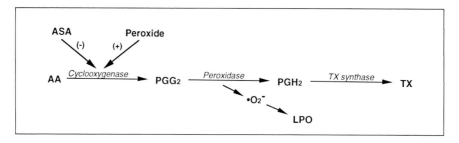

Fig. 1. Mechanism whereby peroxides, by stimulating the activity of the cyclooxygenase enzyme, can increase the synthesis of both thromboxane (TX) and lipid peroxides (LPO), whereas aspirin (ASA), by inhibiting cyclooxygenase, can decrease the synthesis of both. AA = Arachidonic acid; PG = prostaglandin; $\cdot O_2^-$ = oxygen radicals. [Modified from 13.]

The inhibitory effect of low-dose aspirin on thromboxane and lipid peroxide production is manifested in vivo as well as in vitro. In a clinical study of pregnant women at risk of preeclampsia, low-dose aspirin therapy (81 mg/day) resulted in significant reductions in the maternal plasma concentrations of both thromboxane and lipid peroxides [13].

Mechanism for Vasoconstriction: Peroxide Stimulation of Thromboxane

Stimulation of the cyclooxygenase enzyme with exogenous peroxide administration results in increased thromboxane secretion. Thromboxane is a potent vasoconstrictor and one of the most potent vasoconstrictors of the placental vasculature. Therefore, perfusion of the placenta with peroxide results in increased perfusion pressure and increased vascular resistance [14]. Subsequent perfusion of the placenta with aspirin not only blocks the ability of peroxide to increase thromboxane secretion, but also blocks its ability to increase vasoconstriction. Therefore, peroxides per se do not have intrinsic vasoconstrictive properties in the human placenta, but produce vasoconstriction by stimulating the production of thromboxane.

Conclusion

In normal pregnancy there is a gradual favoring of antioxidant activity over peroxidation with advancing gestation. However, in preeclampsia there is an imbalance of increased lipid peroxides and decreased antioxidants. This imbalance might explain the major clinical and pathophysiologic features of preeclampsia, including the imbalance of increased thromboxane (by stimulation of cyclooxygenase) and decreased prostacyclin (by inhibition of prostacyclin synthase) (fig. 2).

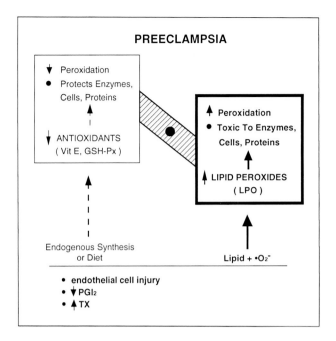

Fig. 2. In preeclampsia there is an imbalance of increased lipid peroxides and decreased vitamin E (Vit E). Boldface type and box for lipid peroxides suggest exacerbation of their actions in preeclampsia, whereas lighter type and box for antioxidants suggest diminution of their actions. PGI_2 = Prostacyclin; TX = thromboxane; $\cdot O_2^-$ = oxygen radicals. [Modified from 8.]

References

1 Ishihara M: Studies on lipoperoxide of normal pregnant women and of patients with toxemia of pregnancy. Clin Chim Acta 1978;84:1–9.
2 Wickens D, Wilkins MH, Lunec J, et al: Free-radical oxidation (peroxidation) products in plasma in normal and abnormal pregnancy. Ann Clin Biochem 1981;18:158–162.
3 Maseki M, Nishigaki I, Hagihara M, et al: Lipid peroxide levels and lipid content of serum lipoprotein fractions of pregnant subjects with or without pre-eclampsia. Clin Chim Acta 1981; 115:155–161.
4 Cranfield IM, Gollan JL, White AG, et al: Serum antioxidant activity in normal and abnormal subjects. Ann Clin Biochem 1979;16:299–306.
5 Jagadeesan V, Prema K: Plasma tocopherol and lipid levels in pregnancy and oral contraceptive users. Br J Obstet Gynaecol 1980;87:903–907.
6 Wang Y, Walsh SW, Guo J, et al: Maternal levels of prostacyclin, thromboxane, vitamin E, and lipid peroxides throughout normal pregnancy. Am J Obstet Gynecol 1991;165:1690–1694.
7 Davidge ST, Hubel CA, Brayden RD, et al: Sera antioxidant activity in uncomplicated and preeclamptic pregnancies. Obstet Gynecol 1992;79:897–901.
8 Wang Y, Walsh SW, Guo J, et al: The imbalance between thromboxane and prostacyclin in preeclampsia is associated with an imbalance between lipid peroxides and vitamin E in maternal blood. Am J Obstet Gynecol 1991;165:1695–1700.
9 Wang Y, Walsh SW, Kay HH: Placental lipid peroxides and thromboxane are increased and prostacyclin is decreased in women with preeclampsia. Am J Obstet Gynecol 1992;167:946–949.

10   Jendryczko A, Drozdz M: Plasma retinol, β-carotene and vitamin E levels in relation to the future risk of preeclampsia. Zentralbl Gynäkol 1989;111:1121–1123.
11   Walsh SW, Wang Y: Secretion of lipid peroxides by the human placenta. Am J Obstet Gynecol 1993;169:1462–1466.
12   Walsh SW: Preeclampsia: An imbalance in placental prostacyclin and thromboxane production. Am J Obstet Gynecol 1985;152:335–340.
13   Walsh SW, Wang Y, Kay HH, et al: Low-dose aspirin inhibits lipid peroxides and thromboxane but not prostacyclin in pregnant women. Am J Obstet Gynecol 1992;167:926–930.
14   Walsh SW, Wang Y, Jesse R: Peroxide induces vasoconstriction in the human placenta by stimulating thromboxane. Am J Obstet Gynecol 1993;169:1007–1012.

Scott W. Walsh, PhD, Department of OB/GYN, Medical College of Virginia,
PO Box 980034, Richmond, VA 23298–0034 (USA)

Galli C, Simopoulos AP, Tremoli E (eds): Effects of Fatty Acids and Lipids in
Health and Disease. World Rev Nutr Diet. Basel, Karger, 1994, vol 76, pp 119–121

....

Does the Long Chain Polyunsaturated Fatty Acid (LCPUFA) Status at Birth Affect the Postnatal LCPUFA Status?

M.M.H.P. Foreman-v. Drongelen[a], *A.C.v. Houwelingen*[a],
C. Koopman-Esseboom[b], *A.D.M. Kester*[c], *P.J.J. Sauer*[b], *G. Hornstra*[a]

Departments of [a]Human Biology and [c]Statistics, University of Limburg, Maastricht;
[b]Sophia Children's Hospital, Rotterdam, The Netherlands

Long chain polyunsaturated fatty acids (LCPUFAs) are of major importance in fetal and infant development. They are preferably used by the body as so-called structural lipids. Two LCPUFAs are particularly important, since they are present in large amounts in the membrane phospholipids of the brain and the retina: these are docosahexaenoic acid (DHA, 22:6ω3) and arachidonic acid (AA, 20:4ω6).

The dietary influence on the postnatal courses of 22:6ω3 and 20:4ω6 has been extensively studied, both in preterm and full-term infants. Infants, particularly those born prematurely, receiving an artificial formula as their main source of dietary lipids, were demonstrated to have a poorer LCPUFA status than infants raised on their own mother's breast milk [1, 2]. While small but significant amounts of 22:6ω3, 20:4ω6 and other LCPUFAs are present in human milk, the majority of the commercially available artificial formulae do not or hardly contain any LCPUFAs [3]. Adding LCPUFAs to artificial formulae proved to alleviate differences between the LCPUFA status of formula- and breast-fed infants [2].

In contrast, a possible influence of the prenatal LCPUFA supply to the fetus in utero on the postnatal LCPUFA status has been given much less attention. Recently, it has been observed that the LCPUFA status at birth, determined in plasma phosphatidylcholine of preterm infants, is strongly correlated to that at the expected date of delivery (EDD), indicating that the experience of the fetus inside the uterus is setting the pattern for what happens postnatally [4]. We tested this hypothesis in a group of preterm infants and in a group of infants born after full-term pregnancy.

Materials and Methods

At birth, a venous cord blood sample and a piece of umbilical cord were collected to provide information on the LCPUFA status at birth. During the postnatal follow-up period, the infants received their own mother's breast milk or an artificial formula as their main source of dietary lipids. At the end of the follow-up period, a neonatal blood sample was taken from the preterm infants at the EDD (± 40 weeks gestational age (GA)) and from the full-term infants at approximately 3 months postnatal age. From the preterm infants, absolute (mg/l) and relative (% of total fatty acids) fatty acid data of both plasma and red blood cell (RBC) phospholipids could be studied. From the term infants, only RBCs were available. The 22:6ω3 and 20:4ω6 values at EDD and 3 months postnatal age were evaluated using multiple regression analysis. As independent variables, the various parameters reflecting the neonatal LCPUFA status at birth as well as the type of postnatal diet (human milk or artificial formula) were used.

Results and Discussion

The variables assessed throughout the study are shown in table 1.

Preterm Infants

From the preterm infants, 25 complete sets of data were available. The postnatal diet was the significant explanatory variable for absolute and relative plasma levels of 22:6ω3 and the relative plasma level of 20:4ω6 at EDD ($p < 0.005$), with lower values in the infants raised on formula. However, the absolute 20:4ω6 level in plasma phospholipids at birth was the only significant explanatory variable (positive correlation, $p < 0.05$) for the 20:4ω6 level in plasma at EDD. For the 22:6ω3 amounts in RBCs at EDD, absolute (mg/kg dry weight) and relative 22:6ω3 amounts in arterial umbilical vessel walls were the only and highly explanatory variable (positive correlation, $p \leq 0.001$). Since the LCPUFA compositions of umbilical vessel walls are considered to be a long-term reflection of the fetal LCPUFA status in utero [5], this finding strongly supports the hypothesis that the LCPUFA status at birth is a major determinant of the postnatal LCPUFA status.

Full-Term Infants

In the term children, the results (in 37 complete sets of variables) were less decisive than in the preterm group. Both the postnatal diet ($p < 0.02$, with again lower levels in the subgroup fed artificial formula) and the relative amount of 22:6ω3 in the venous cord vessel walls (positive correlation, $p < 0.05$) added significantly to the explanation of the relative amount of 22:6ω3 at 3 months postnatal age. For the absolute amount of 20:4ω6 in RBCs at 3 months postnatal age, the amount of 20:4ω6 in the arterial cord vessel walls was the only significant explanatory variable ($p < 0.01$).

Table 1. Explanatory variables for postnatal LCPUFA status

| | Preterm infants | | | | Term infants | |
| | plasma PLs | | RBC PLs | | RBC PLs | |
	mg/l	%	mg/l	%	mg/l	%
22:6ω3 all	diet	diet	UA	UA	–	diet
22:6ω3 bw	diet	diet	UA	UA	–	diet
						UV
20:4ω6 all	CB	(diet)	–	–	(UA)	–
20:4ω6 bw	CB	diet	–	–	UA	–

Regression analysis: all = total regression equation calculated; bw = backward analysis. LCPUFA status at birth: CB = cord blood; UA = umbilical artery; UV = umbilical vein. All variables: $p < 0.05$, except for (diet) resp. (UA): $p < 0.10$.

Conclusions

In conclusion, our results demonstrate an influence of both the postnatal (the diet) and the prenatal LCPUFA supply (reflected in the LCPUFA status at birth) on the postnatal LCPUFA status. Consequently, both the postnatal and the prenatal LCPUFA supply need to be optimized to ensure an optimal postnatal LCPUFA status, especially for children born prematurely. This implies an adequate maternal diet during pregnancy, providing a proper LCPUFA supply to the fetus growing in utero.

References

1   Carlson SE, Rhodes PG, Ferguson MG: Docosahexaenoic acid status of preterm infants at birth and following feeding with human milk or formula. Am J Clin Nutr 1986;44:798–804.
2   Clandinin MT, Parrott A, Van Aerde JE, et al: Feeding preterm infants a formula containing C_{20} and C_{22} fatty acids simulates plasma phospholipid fatty acid composition of infants fed human milk. Early Hum Dev 1992;31:41–51.
3   Jensen RG, Ferris AM, Lammi-Keefe CJ: Lipids in human milk and infant formulas. Annu Rev Nutr 1992;12:417–441.
4   Leaf AA, Leighfield MJ, Ghebremeskel K, et al: Polyunsaturated fatty acids in plasma choline phosphoglycerides of preterm infants (abstract). 3rd Int Congr Essential Fatty Acids and Eicosanoids, Adelaide, 1992, p 77.
5   Hornstra G, v. Houwelingen AC, Simonis M: Fatty acid composition of umbilical arteries and veins: Possible implications for the fetal EFA status. Lipids 1989;24:511–517.

M. M. H. P. Foreman-v. Drongelen, MD, Department of Human Biology, University of Limburg, PO Box 616, NL–6200 MD Maastricht (The Netherlands)

Galli C, Simopoulos AP, Tremoli E (eds): Effects of Fatty Acids and Lipids in
Health and Disease. World Rev Nutr Diet. Basel, Karger, 1994, vol 76, pp 122–125

....

Effects of Fish Oil Supplementation in Late Pregnancy on Prostaglandin Metabolism[1]

Jannie Dalby Sørensen, Sjúrdur F. Olsen

Perinatal Epidemiologic Research Unit, Department of Obstetrics and Gynecology,
Aarhus University Hospital, and Institute of Epidemiology and Social Medicine,
University of Aarhus, Denmark

Fish Oil and Prostaglandin Metabolism

Marine food sources are rich in long-chain polyunsaturated ω3 fatty acids [1, 2]. Together the ω3 and ω6 fatty acids constitute the essential fatty acids, which means that they need to be provided through the diet to avoid deficiency symptoms. The main long-chain ω3 fatty acids are α-linolenic acid (LNA), eicosapentaenoic acid (EPA), docosapentaenoic acid (DPA), and docosahexaenoic acid (DHA), of which the latter three are derived from marine food sources. The main ω6 fatty acids are linoleic acid (LA) and arachidonic acid (AA).

The essential fatty acids are used for cell structures and their oxidative products, the eicosanoids, as local hormone-like regulators of cellular function. The enzyme systems for metabolism of fatty acids are shared and different fatty acids are competing with each other. AA is usually regarded as the main prostaglandin (PG) precursor and gives rise to dienoic PGs, including prostacyclin (PGI_2), thromboxane A_2, PGE_2 and $PGF_{2\alpha}$. The ω3 fatty acids give rise to trienoic counterparts, which are of low biological activity except for PGI_3, which seems equipotent to the AA-derived counterpart. It is known from studies of nonpregnant subjects that when intake of ω3 fatty acids increases, the formation of TxA_2, PGE_2, and $PGF_{2\alpha}$ decreases, but for some reason PGI_2

[1] The results cited from prostaglandin analyses were obtained in cooperation with Garret A. FitzGerald, Dublin. The study was supported by Sygekassernes Helsefond, Direktør Jacob Madsen og Hustru Olga Madsens Fond and Lida og Oskar Nielsens Fond.

remains unchanged. The formation of the trienoic prostanoids increases as well. These effects affecting thromboxane/prostacyclin ratio are expected to cause vasodilation and to be antiaggregatory. Other effects of ω3 fatty acids with circulatory consequences have been described in nonpregnant subjects, including a reduction in blood viscosity, a lower blood pressure and a decreased vasospastic response to angiotensin II and catecholamines.

Prostaglandins in Pregnancy

Pregnancy and parturition are characterized by extensive changes in PG metabolism. Of the large PG family, it is compounds produced through the metabolism of AA via the cyclooxygenase pathway that have been most extensively studied in pregnancy; these include PGs of the D, E or F series, prostacyclin, and its endogenous antagonist, thromboxane A_2. Compared to nonpregnant subjects there is a predominance of prostacyclin relative to thromboxane A_2 in pregnancy, which contributes to the adaptational changes of the systemic and uteroplacental circulations in pregnancy and may also be of significance for controlling myometrial activity [3]. In pregnancy-induced hypertension, preeclampsia and intrauterine growth retardation these adaptational changes are inadequate probably as a result of an insufficient increase in prostacyclin production [4].

The above mentioned effects of ω3 fatty acids on thromboxane/prostacyclin ratio, and the other effects influencing circulatory function, could be beneficial in pregnancy and especially in pregnancy complicated by hypertension or preeclampsia, since the resulting effect of ω3 fatty acids would be expected to improve placental circulation, which is reduced in these pathological pregnancies [5–7].

Because of these perspectives, and because any extrapolation from findings obtained in studies of nonpregnant subjects are highly uncertain due to the mentioned extensive changes in PG metabolism induced by pregnancy, we found it relevant to ask the question: Will marine ω3 fatty acids, provided in doses that can be applied in prophylactic studies, influence PG production in human pregnancy in a way that can be expected to be of value in pregnancy-induced hypertension and preeclampsia?

Outline of Study

We have conducted in Aarhus a controlled intervention study of the effects of ω3 fatty acid supplementation in normal pregnancy on pregnancy duration, birthweight, blood pressure and, in subgroups, on maternal and fetal PG production, essential fatty acid status, coagulation and fibrinolysis factors, and fatty acid content of human milk. 533 healthy pregnant women were randomized in the 30th week of pregnancy to receive fish oil, olive oil or no oil supplementation. The fish oil supplementation consisted of four 1-gram capsules/day with fish oil corresponding to 2.7 g of ω3 fatty acids. One of the control groups received four 1-gram olive oil capsules, the other received no supplementation [for details, see 8]. Forty-seven women entered the subgroup in which biochemical analyses of PG metabolites were performed.

We found an increased production of the trienoic prostanoids (PGI_3-M and TxB_3) in the 37th week, i.e. after 7 weeks of supplementation, in the fish oil group compared to the combined control group. Contrary to expectation from studies in nonpregnant subjects, however, we could find no effects on the maternal TxB_2 formation in the fish oil group. In the umbilical cord, on the other hand, a lower concentration of TxB_2 was found in the fish oil group [9].

These data are currently being explored further for associations between intake of ω3 fatty acids and prostanoid formation, using an approach other than the earlier one which was simply to compare prostanoid levels in the groups as defined by random assignment to either fish oil or control treatments. The idea is to use the levels of ω3 fatty acids in plasma phospholipids as a measure for the ω3 fatty acid intake at the individual level [10], which not only will reflect group assignment but also how well the individual complied to the treatment regimen and her background intake of ω3 fatty acids.

Also the possible prophylactic and therapeutic effect of ω3 fatty acids against pregnancy-induced hypertensive disorders, intrauterine growth retardation, and preterm birth, is currently being intensively studied in a multicenter trial organized in Aarhus, Denmark.

References

1   Leaf A, Weber P: Cardiovascular effects of n-3 fatty acids. N Engl J Med 1988;318:549–557.
2   Hansen HS: Dietary essential fatty acids and prostaglandin formation in vivo; in Taylor TG, Jenkins NK (eds): Proc 13th Int Congr Nutrition, 1985. London, Libbey, 1986, p 353.
3   Noort WA, Keirse MJNC: Prostacyclin versus thromboxane metabolite excretion: Changes in pregnancy and labour. Eur J Obstet Gynecol Reprod Biol 1990;35:15–21.
4   Friedman SA: Preeclampsia: A review of the role of the prostaglandins. Obstet Gynecol 1988; 71:122–137.
5   Dyerberg J, Bang HO: Preeclampsia and prostaglandins (letter). Lancet 1985;i:1267.
6   England MJ, Atkinson PM, Sonnendecker EWW: Pregnancy-induced hypertension: Will treatment with dietary eicosapentaenoic acid be effective? Med Hypotheses 1987;24:179–186.
7   Secher NJ, Olsen SF: Fish oil and pre-eclampsia. Br J Obstet Gynaecol 1990;97:1077–1079.
8   Olsen SF, Sørensen JD, Secher NJ, et al: Does fish oil supplementation in late pregnancy prolong pregnancy? A randomised controlled trial. Lancet 1992;339:1003–1007.

9   Sørensen JD, Olsen SF, Pedersen AK, Boris J, Secher NJ, FitzGerald GA: Effects of fish oil
    supplementation in the third trimester of pregnancy on prostacyclin and thromboxane production.
    Am J Obstet Gynecol 1993;168:915–922.
10 Houwelingen AC v., Hornstra G, Kromhout D, de Lezenne Coulander C: Habitual fish consump-
    tion, fatty acids of serum phospholipids and platelet function. Atherosclerosis 1989; 75:157–165.

Dr. Jannie Dalby Sørensen, Perinatal Epidemiologic Research Unit, University of Aarhus,
Nørrebrogade 37–39, DK–8000 Aarhus C (Denmark)

Galli C, Simopoulos AP, Tremoli E (eds): Effects of Fatty Acids and Lipids in
Health and Disease. World Rev Nutr Diet. Basel, Karger, 1994, vol 76, pp 126–129

....

Plasma Total Phospholipid Arachidonic Acid and Eicosapentaenoic Acid in Normal and Hypertensive Pregnancy

M.C. Craig-Schmidt [a], *S.E. Carlson* [b], *L. Crocker* [b], *B.M. Sibai* [b]

[a] Department of Nutrition and Food Science, Auburn University, Auburn, Ala.;
[b] Departments of Pediatrics and Obstetrics and Gynecology, University of Tennessee,
Memphis, Tenn., USA

Several investigators [1–3] have postulated that preeclampsia is due to imbalances in arachidonic acid (AA) metabolites. In particular, an increased ratio of thromboxane A_2 to prostacyclin has been suggested as a mechanism for some of the pathologic changes seen in preeclampsia. Prostacyclin, a potent vasodilator, is decreased in preeclampsia compared to normal pregnancies [4–6]. In part, this decrease may be due to inhibition of prostacyclin synthetase by increased concentrations of lipoxygenase products, such as hydroperoxy-eicosatetraenoic acid (HPETE) and hydroxyeicosatetraenoic acid (HETE) [1]. In addition, elevated thromboxane A_2 has been observed in patients with pregnancy-induced hypertension [7], shifting the eicosanoid balance further toward vasoconstriction. Also, increased concentrations of leukotrienes in preeclampsia may explain the edema and proteinuria seen in this condition. Because AA and its eicosanoid metabolites appear to play a role in preeclampsia, the present study was undertaken to determine if there were inherent differences in the ω6 and ω3 fatty acid status of women with normotensive pregnancies compared to that of women with hypertensive disorders of pregnancy. The fatty acid composition of plasma phospholipids in control pregnant subjects was compared to that of women with gestational hypertension, preeclampsia, and chronic hypertension.

Table 1. Criteria for subject selection

	High blood pressure[1]	Proteinuria[2]
NORM	–	–
GH	+	–
PEC	+	+
CH	+ (preexisting)	+/–

[1] High blood pressure: $>140/90$.
[2] Proteinuria: >300 mg/24 h.

Methods

The subjects were nulliparous women, 21 ± 6 (mean $\pm$ SD) years of age. Thirty-six subjects were divided into four experimental groups (table 1). Hypertension was defined as blood pressure of greater than 140/90 measured on at least two occasions 6 h apart; proteinuria was defined as greater than 300 mg urinary protein/24 h. Ten subjects had normotensive pregnancies (NORM). Ten subjects with gestational hypertension (GH) exhibited high blood pressure but no proteinuria, whereas 10 subjects with preeclampsia (PEC) had both high blood pressure and proteinuria. Subjects (n = 6) assigned to the chronic hypertension (CH) group had high blood pressure prior to pregnancy, and some of these patients exhibited proteinuria. Venous blood samples were drawn late in the third trimester of pregnancy at 36 ± 1 (mean $\pm$ SD) weeks of gestation. After extraction of lipids from the plasma, the total phospholipids and individual classes were isolated by thin-layer chromatography and individual fatty acids separated by capillary gas chromatography as described by Carlson et al. [8].

Results

In the fatty acid profile of plasma phospholipids (table 2), no differences were observed among groups in the concentrations of saturated, monounsaturated, or polyunsaturated fatty acids. Similarly, there were no differences in either the content of $\omega6$ or the $\omega3$ fatty acids between women with uncomplicated pregnancies and those with gestational hypertension, preeclampsia, or chronic hypertension.

Plasma phospholipid AA concentrations (table 3) did not differ among the normotensive, gestational hypertensive, and preeclamptic groups. The women with chronic hypertension, however, had significantly greater concentrations of AA in the plasma phospholipids in comparison with the other three groups. This was true for AA values expressed as concentration in mg/l or as percent of total fatty acids in plasma phospholipids. When plasma phospholipids were

Table 2. Fatty acid profile of plasma total phospholipids from normal and hypertensive pregnant women[1]

	NORM, mg/l	GH, mg/l	PEC, mg/l	CH, mg/l
Saturated	441 ± 171	492 ± 120	556 ± 122	576 ± 178
Monounsaturated	162 ± 67	182 ± 38	206 ± 57	190 ± 59
Polyunsaturated	505 ± 109	539 ± 155	557 ± 104	609 ± 148
ω6	453 ± 98	479 ± 141	503 ± 96	546 ± 132
ω3	52 ± 12	59 ± 16	54 ± 10	63 ± 16

[1] Values are means ± SD. No differences were observed among groups ($p > 0.05$).

Table 3. AA and EPA in plasma total phospholipids from normal and hypertensive pregnant women[1]

	NORM	GH	PEC	CH
AA, mg/l	134 ± 30[a]	130 ± 32[a]	141 ± 24[a]	176 ± 45[b]
AA, % total FA	11.2 ± 1.3[a]	10.7 ± 1.8[a]	10.7 ± 1.3[a]	12.8 ± 0.9[b]
EPA, mg/l	2.8 ± 1.6	2.8 ± 1.6	2.9 ± 0.8	3.2 ± 1.8
AA/EPA ratio	69 ± 45	55 ± 16	54 ± 19	70 ± 31
ω6/ω3 ratio	8.8 ± 1.2	8.1 ± 1.1	9.4 ± 1.3	8.8 ± 0.7

[1] Values are means ± SD. Means in the same row with different superscripts differ (ANOVA; Fisher PLSD significant at 95%).

further fractionated into various classes, AA was significantly greater in the phosphatidylcholine fraction of the group with chronic hypertension (131 ± 28 mg/l) in comparison to the other groups. No differences in phosphatidylcholine arachidonate were observed among the normotensive (104 ± 21 mg/l), gestational hypertensive (94 ± 20 mg/l) and preeclamptic subjects (107 ± 17 mg/l).

Because there is interest in treating hypertensive disorders of pregnancy with fish oil supplementation as a means of restoring eicosanoid balance, baseline concentrations of eicosapentaenoic acid (EPA) also were determined in these subjects. No differences in plasma phospholipid EPA concentrations were observed among any of the groups nor were differences seen in the AA/EPA ratio or the ω6/ω3 ratio (table 3).

Conclusions

Differences in AA and EPA content or their balance were not present in preeclampsia or gestational hypertension compared to normotensive pregnancy, even late in the third trimester. Therefore, differences in ω6 and ω3 fatty acids in plasma phospholipids of pregnant women with preeclampsia and gestational hypertension versus normal pregnant women canot be primarily responsible for these hypertensive disorders.

In pregnant women with chronic hypertension, however, greater concentrations of AA in plasma phospholipids were observed in comparison with normal pregnant women. This increase in phospholipid AA is consistent with the model proposed by Ogburn et al. [1] who observed an increase in phospholipid AA in maternal blood of preeclamptic women, possibly as a result of a shift from nonesterified and triglyceride AA in cord blood. The results obtained in the present study for preeclampsia and gestational hypertension, however, do not fit this model.

Although no differences were observed in EPA concentrations among groups in the present experiment, diets with fish oil would be expected to alter both plasma and tissue AA and EPA concentrations and balance. Supplementation with fish oil as a source of EPA could have a secondary influence on the balance of eicosanoids produced from AA in women prone to preeclampsia, gestational hypertension or chronic hypertension. The results of this study do not preclude further investigation of fish oil supplementation in women with hypertensive disorders of pregnancy.

References

1   Ogburn PL, Williams PP, Johnson SB, et al: Serum arachidonic acid levels in normal and preeclamptic pregnancies. Am J Obstet Gynecol 1984;148:5–9.
2   Walsh SW: Preeclampsia: An imbalance in placental prostacyclin and thromboxane production. Am J Obstet Gynecol 1985;152:335–340.
3   Wang Y, Kay HH, Killam AP: Decreased levels of polyunsaturated fatty acids in preeclampsia. Am J Obstet Gynecol 1991;164:812–818.
4   Remuzzi G, Marchesi D, Zoja C, et al: Reduced umbilical and placental vascular prostacyclin in severe pre-eclampsia. Prostaglandins 1980;20:105–110.
5   Goodman RP, Killam AP, Brash AR, et al: Prostacyclin production during pregnancy: Comparison of production during pregnancy and pregnancy complicated by hypertension. Am J Obstet Gynecol 1982;142:817–822.
6   Wang Y, Walsh SW, Guo J, et al: The imbalance between thromboxane and prostacyclin in preeclampsia is associated with an imbalance between lipid peroxides and vitamin E in maternal blood. Am J Obstet Gynecol 1991;165:1695–1700.
7   Fitzgerald DJ, Rocki W, Murray R, et al: Thromboxane A_2 synthesis in pregnancy-induced hypertension. Lancet 1990;335:751–754.
8   Carlson SE, Cooke RJ, Rhodes PG, et al: Long-term feeding of formulas high in linolenic and marine oil to very low birth weight infants: Phospholipid fatty acids. Pediatr Res 1991;30:404–412.

Dr. Margaret Craig-Schmidt, Department of Nutrition and Food Science,
Auburn University, Auburn, AL 36849 (USA)

Galli C, Simopoulos AP, Tremoli E (eds): Effects of Fatty Acids and Lipids in
Health and Disease. World Rev Nutr Diet. Basel, Karger, 1994, vol 76, pp 130–132

....

Summary Statement:
Clinical Trials with ω3 Fatty Acids

Raffaele De Caterina

The session was co-chaired by *A. Leaf* and *R. De Caterina,* and presentations were made by Drs. *De Caterina, J. Dyerberg, D.R. Robinson, R.I. Lorenz,* and *H. Knapp.*

This symposium was aimed at covering the expanding areas of clinical use of ω3 fatty acids in some (but by no means all) human diseases (important areas such as skin diseases, diabetes and cancer were not covered). Stemming from the original observations of Kromann and Green, and, subsequently, by Bang and Dyerberg, reporting the different incidence of a variety of chronic diseases in Greenland Eskimos as compared to control Scandinavian populations, most research emphasis has been put so far in trying to confirm the epidemiological evidence of an inverse relationship of the ω3 fatty acid content in the diet (essentially from marine sources) and the incidence of coronary heart disease (CHD). This also led to intervention trials mostly in this area. The first presentation in the symposium reviewed the epidemiological data for this relationship. Indeed, although not unanimously, the bulk of the epidemiological evidence gathered so far does confirm the idea of an inverse relationship between the dietary content of ω3 fatty acids and the incidence of CHD. Although other dietary components, especially the content of saturated fats, certainly play a major role in influencing the risk of CHD, an independent protective effect of ω3 fatty acids has been found in several cross-cultural comparisons among and within different populations, and such evidence has been defined 'sufficiently substantial' to justify the present efforts to elucidate the mechanisms for such an effect and the performance of controlled clinical studies. Although the main emphasis has hitherto been put on data describing an antithrombotic action, newer epidemiological and experimental data also point to an antiatherosclerotic effect and, possibly, on effects against the functional consequences of myocardial ischemia (i.e. the electrical myocardial instability).

The different disease patterns originally described in Greenland Eskimos suggested other areas of investigation for possible therapeutic effects in human disease. ω3 fatty acids have a number of biological reasons for their use in renal diseases, including, among others, a possible increase in the renal vasodilatory capacity by a rearrangement of renal prostanoid production, a reduction in the production of proinflammatory leukotrienes, a reduction in the transcapillary escape rate of albumin, and actions limiting cyclosporine-related nephrotoxity. Animal models of renal diseases, mostly of immune renal disease, have supported the idea of possible usefulness of these compounds in humans. Most promising areas of investigations at the moment include the reduction of proteinuria in some chronic glomerular diseases, the treatment of IgA nephropathy and the prevention of cyclosporine-induced nephrotoxicity. In these last two areas, the results of two large ongoing multicenter clinical trials are eagerly awaited.

The treatment of human autoimmune disease with ω3 fatty acids was reviewed. Dietary supplements with these substances have been shown to have anti-inflammatory effects on autoimmune diseases in experimental animals and in human subjects. Seven well-controlled double-blind clinical trials of the effects of dietary supplements of ω3 fatty acids in rheumatoid arthritis have reported statistically significant beneficial effects, which were, however, of a small magnitude and modest clinical impact. Such studies were conducted with doses in the range of 5–6 g/day, with minimal side effects, justifying the hypothesis that larger doses might have possibly a greater clinical impact.

Clinical trials with ω3 fatty acids in chronic inflammatory bowel diseases, namely ulcerative colitis and Crohn's disease, in which a long-term treatment with drugs possessing a number of adverse effects is presently required for many patients, were reviewed. The favorable safety profile of ω3 fatty acids adds to their rationale as anti-inflammatory and immunomodulatory agents. Four double-blind randomized placebo-controlled trials in ulcerative colitis with doses of ω3 fatty acids ranging between 2.7 and 5.4 g/day have documented moderate clinical improvements, mostly in remission induction. On the other hand, in Crohn's disease, the experience gathered so far has failed to find significant clinical effects.

The experience with ω3 fatty acids in respiratory diseases, where the attenuation of vaso- and bronchoactive leukotriene production by ω3 fatty acids prompted studies mostly in allergic respiratory diseases, were presented. So far, one study found no change in allergic asthma with ω3 fatty acid supplementation, and another claimed exacerbations of symptoms in aspirin-sensitive asthmatics, attributed to in vivo cyclooxygenase inhibition. In a nasal allergen challenge model in which inhibition of 5-lipoxygenase reduced nasal symptoms, dietary ω3 fatty acids modified the pattern but not the overall production of leukotrienes and did not attenuate symptoms. However, since the late component of respiratory hypersensitivity reactions may be altered favorably by ω3 fatty acids, further research in syndromes of pulmonary inflammation was felt to be warranted.

Galli C, Simopoulos AP, Tremoli E (eds): Effects of Fatty Acids and Lipids in
Health and Disease. World Rev Nutr Diet. Basel, Karger, 1994, vol 76, pp 133–136

....

The Epidemiology of ω3 Fatty Acids

J. Dyerberg

Medi-Lab A.S., Copenhagen, Denmark

Our observations in Greenland Eskimos suggest that the high intake of ω3 polyunsaturated fatty acids (PUFA) of the Eskimos was associated with their low mortality rate from coronary heart disease (CHD) [1, 2]. Other epidemiological observations supported these studies [3, 4]. However, many other cross-cultural differences than the consumption of ω3 PUFA exist between these populations. In this respect a series of within-population studies is of great interest.

Fatty Acids and CHD

In the 1960s and 70s the interest of research was focused on the ω6 PUFA [5–9]. The 'Eskimo experience' [1, 2, 10, 11] has, however, shifted a major part of the present epidemiological interest in PUFA towards the ω3 family. The Seven Countries Study showed a 7-fold difference in average saturated fat intake between the 16 cohorts involved in that subset of analysis [12, 13]. The average saturated fat intake was found to be strongly related to 15 years' mortality from CHD. These results suggest that populations characterized by a high saturated fat intake have higher mortality from CHD.

In the Framingham Heart Study, the Honolulu Heart Study, and the Puerto Rico Heart Help Programme, no association was found between the intake of different fatty acids and CHD incidence [14]. This was also true for the Zutphen Study [15], whereas in the Western Electric Study a significant inverse association was found between PUFA intake in CHD mortality [16].

The Greenland Eskimo Experience

Our studies in the 1970s [1, 2] showed that the average intake of ω3 PUFA in Eskimos was 2–3 times higher than the intake of ω6 PUFA. The diet of the Eskimos was characterized by a high fat content (39% of energy from fat) with a low amount of saturated fat, and a high polyunsaturate/saturate (P/S) ratio of 0.84. Also the intake of monounsaturated fatty acids was very high. In 1980, Kromann and Green [17] published the results of a survey of several chronic diseases in the Upernavik district in North-West Greenland. They showed that in comparison with Danish mortality data, only 3 cases of acute myocardial infarction were found, while 40 cases were expected, based on the Danish study. Since then, these findings were further substantiated by epidemiological surveys of overall Greenland [18]. Even if the Eskimo data created great interest in ω3 PUFA, the data are not sufficiently valid to justify a conclusion of a cause-relationship to ischemic heart disease.

Other Epidemiological ω3 Studies

In a cross-cultural study from 21 countries, the relation between fish consumption and CHD was investigated [3]. A moderate negative association was found which appeared stable over different periods. The correlation was, however, very much dependent on the Japanese data. Exclusion of the Japanese data substantially reduced the correlation between fish consumption and CHD. The completion of the mortality data of the Seven Countries Study now allows for analysis relating fish consumption data collected at baseline to 15 years' CHD mortality data [19]. The study only reveals a weak, nonsignificant correlation coefficient between fish consumption at baseline and 15 years' mortality from CHD.

The detailed analysis of the Zutphen Study from Holland included in the Seven Countries Study allows for a more detailed within-cohort analysis of the relation between fish consumption and CHD mortality [20]. It showed that fish consumption at the start of the study was inversely related to mortality from CHD in the period 10–20 years later. Mortality from CHD was about 2 times lower among males eating at least 30 g of fish/day compared with noneaters. Dose/response relation was observed between 0 and 30 g of fish/day.

In the 1957 Western Electric Study, a cohort study conducted in 2,000 males in Chicago, an inverse relation was also found between fish consumption in 1957 and mortality of CHD during 25 years of follow-up [21]. The same result came from a 14-year follow-up study of approximately 11,000 Swedes [22]. This was, however, not the case in two cohort studies carried out in

Hawaii [23] and Norway [24], in which no relation was found between fish consumption and CHD.

In the large American MRFIT Study, dietary fish intake in the usual care group comprising more than 6,000 males aged 35–57 years, has been recorded, and the group has now been followed for more than 10 years. The daily intake of ω3 PUFA has recently been reported to be inversely related to total mortality and mortality from ischemic heart disease [25].

Of present major interest are the epidemiological findings in Alaskan Eskimos showing a substantially lower extent of atherosclerosis as compared to nonnative Alaskans [26]. Also, the hitherto unpublished results from an ongoing trial in quantifying the extent of atherosclerosis in Alaskan and Greenland Eskimos point in the same direction.

References

1   Bang HO, Dyerberg J, Hjorne N: The composition of food consumed by Greenland Eskimos. Acta Med Scand 1976;200:69–73.
2   Bang HO, Dyerberg J, Sinclair HM: The composition of the Eskimo food in North Western Greenland. Am J Clin Nutr 1980;33:2657–2661.
3   Crombie IK, Mcloone P, Smith WCS, Thomson M, Tunstall Pedoe H: International differences in coronary heart disease mortality and consumption of fish and other foodstuffs. Eur Heart J 1987; 8:560–563.
4   Hirai A, Hamazaki T, Terano T, Nishikawa T, Tamura Y, Kumagai A: Eicosapentaenoic acid and platelet function in Japanese. Lancet 1980;ii:1132–1133.
5   Hagve T-A, Christensen E, Groenn M, Christophersen BO: Regulation of the metabolism of polyunsaturated fatty acids. Scand J Clin Lab Invest 1988;48(suppl 191):33–46.
6   Kinsell LW, Partridge J, Boling L, Marven S, Michaels GP: Dietary modification of serum cholesterol and phospholipid levels. J Clin Endocrinol 1952;12:909–913.
7   Ahrens EH, Blankenhorn DH, Tsaltas TT: Effect on human serum lipids of substituting plant for animal fat in the diet. Proc Soc Exp Biol Med 1954;86:672–678.
8   Keys A, Anderson TT, Grande F: Prediction of serum cholesterol responses of man to change in fats in the diet. Lancet 1957;ii:959–966.
9   Goodnight SH Jr, Harris WS, Connor WE, Illingworth PR: Polyunsaturated fatty acids, hyperlipidemia and thrombosis. Arteriosclerosis 1982;2:87–113.
10 Bang HO, Dyerberg J: The lipid metabolism in Greenlanders. Meddr Groenland, Man Soc 1981; 2:3–18.
11 Dyerberg J, Bang HO, Stoffersen HO, Moncada S, Vane J: Eicosapentaenoic acid and prevention of thrombosis and atherosclerosis. Lancet 1978;ii:117–119.
12 Keys A: Coronary heart disease in seven countries. Circulation 1970;41(suppl 1):1–211.
13 Keys A, Menotti A, Karvonen MJ, et al: The diet and 15-year death rate in the Seven Countries Study. Am J Epidemiol 1986;124:903–915.
14 Gordon T, Kagan A, Garcia-Palmeri M, et al: Diet and its relation to coronary heart disease and death in three populations. Circulation 1981;63:500–515.
15 Kromhout D, de Lezenne Coulander C: Diet, prevalence and 10-year mortality from coronary heart disease in 871 middle-aged men. The Zutphen Study. Am J Epidemiol 1984;119:733–741.
16 Shekelle RB, Shryock AM, Paul O, et al: Diet, serum cholesterol, and death from coronary heart disease. The Western Electric Study. N Engl J Med 1981;304:65–70.
17 Kromann N, Green A: Epidemiological studies in the Upernavik district, Greenland. Acta Med Scand 1980;208:401–406.
18 Bjerregaard P, Dyerberg J: Mortality from ischaemic heart disease and cerebrovascular disease in Greenland. Int J Epidemiol 1988;17:514–519.

19   Kromhout D, Keys A, Aravanis C, et al: Food consumption patterns in the nineteen sixties in seven countries. Am J Clin Nutr 1989;49:889–894.
20   Kromhout D, Bosschieter EB, De Lezenne Coulander C: The inverse relation between fish consumption and 20-year mortality from coronary heart disease. N Engl J Med 1985;312:1205–1209.
21   Shekelle R, Missell L, Paul O, MacMillan-Schryock A, Stamler J: Fish consumption and mortality from coronary heart disease (letter). N Engl J Med 1985;313:820.
22   Norell SE, Ahlbom A, Feychting M, Pedersen NL: Fish consumption and mortality from heart disease. Br Med J 1986;293:426.
23   Curb JD, Reed DM: Fish consumption and mortality from coronary heart disease (letter). N Engl J Med 1985;313:821.
24   Vollset SE, Heuch I, Bjelke E: Fish consumption and mortality from coronary heart disease. N Engl J Med 1985;313:820–821.
25   Dolecek TA, Grandits G: Dietary polyunsaturated fatty acids and mortality in the Multiple Risk Factor Intervention Trial (MRFIT); in Simopoulos AP, Kifer RR, Martin RE, Barlow SM (eds): Health Effects of ω3 Polyunsaturated Fatty Acids in Seafoods. World Rev Nutr Diet. Basel, Karger, 1991, vol 66, pp 205–216.
26   Middaugh JP: Cardiovascular deaths among Alaskan natives 1980–86. Am J Public Health 1990; 80:282–285.

J. Dyerberg, MD, Medi-Lab AS, 7, Adelgade, DK–1304 Copenhagen K (Denmark)

Galli C, Simopoulos AP, Tremoli E (eds): Effects of Fatty Acids and Lipids in
Health and Disease. World Rev Nutr Diet. Basel, Karger, 1994, vol 76, pp 137–142

....

ω3 Fatty Acids in Renal Diseases

Raffaele De Caterina

CNR Institute of Clinical Physiology, Pisa, Italy, and Cardiovascular Division,
Brigham and Women's Hospital, Harvard Medical School, Boston, Mass., USA

Attention as to possible influences of ω3 polyunsaturated fatty acids
(PUFA) on human renal diseases has until now been relatively scarce, but
evidence has started to accumulate about their efficacy and potential clinical
usefulness. This brief review aims to summarize the main studies of clinical
efficacy of ω3 PUFA in renal disease. The biological rationales for their use in
this setting are summarized in table 1. These rationales are backed up by a
number of animal studies in models of immune renal diseases, reduced renal
mass and other renal diseases, the coverage of which is beyond the scope of this
work.

Studies in Humans

Three patient categories with renal diseases have been predominantly
studied until now in the search for possible benefits from ω3 PUFA supplementation: (1) renal patients with hyperlipidemia; (2) patients with renal toxicity
related to the use of cyclosporine; and (3) patients with immune-related renal
diseases.

Studies in Renal Patients with Hyperlipidemia

In chronic hemodialysis patients, in whom accelerated atherosclerosis is a
frequent finding, ω3 PUFA have been proven effective in decreasing serum

Table 1. Reasons for use of ω3 fatty acids in renal disease

Effects on blood lipids (plasma triglycerides and VLDL)
Alterations in mediators of the immune response (down-regulation of the immune system in immune renal disease)
Alterations of renal eicosanoid production, possibly increasing renal vasodilatory reserve
Reduction in renal TXA_2 and $PGF_{2\alpha}$, biologically unbalanced by the increase in TXA_3 and $PGF_{3\alpha}$
Parallel reduction of renal PGI_2 and PGE_2 and increase in PGI_3 and PGE_3
Reduction in bioactive leukotriene production by inflammatory cells
Alterations of blood rheology
Decrease in blood pressure
Decrease in transcapillary escape rate of albumin
Decrease in platelet aggregability
Alteration of renal vascular reactivity to angiotensin II
Decrease in efferent arteriolar resistance
Antagonism (by multiple mechanisms) of cyclosporine toxicity

triglycerides, total cholesterol and blood pressure in hemodialysis patients with different dyslipidemias (types IIa, IIb and IV) in three studies [1–3]; serum triglycerides and blood pressure in another [4]; serum triglycerides and the ApoB/ApoA1 ratio in another [5]. Similar findings with respect to triglycerides have been found in two studies in patients in chronic ambulatory peritoneal dialysis [6, 7], although divergent results were found regarding total cholesterol and high density lipoprotein (HDL)-cholesterol, changing favorably in one study [6] and unfavorably in the other [7]. A decrease in triglyceride (but not cholesterol) concentrations has also been reported in patients with nephrotic syndrome [8], in whom dyslipidemia – especially hypertriglyceridemia – is common and may explain the frequent association with coronary heart disease and its complications. Thus, it appears that kidney patients may benefit from ω3 PUFA with regard to a decrease in plasma triglyceride concentrations, as reported in other patient categories and in healthy subjects.

Cyclosporine-Induced Toxicity

Several reports have appeared regarding the use of ω3 PUFA in patients to prevent cyclosporine-induced nephrotoxicity, a condition which is associated with increased renal production of thromboxane (TX) [9], but which could be affected by ω3 PUFA also via other mechanisms (for instance via their hypotensive effect). According to the results of one study, actually, some preservation of renal function, evaluated by estimates of the slope of reciprocal plasma creatinine versus time, was present in 14 adult renal transplant recipients with chronic vascular rejection, in which the majority (9 cases) did not receive cyclosporine; this suggests some possible interferences with general mechanisms of transplant rejection [10]. However, most of the available data suggest some effects on cyclosporine-related nephrotoxicity. In one study patients with psoriasis and receiving cyclosporine plus ω3 PUFA were compared to a group of patients receiving cyclosporine alone. The cyclosporine-induced decrease in glomerular filtration rate was attenuated and the decrease in renal plasma flow completely prevented by ω3 PUFA [11]. The same holds true for postoperative, cyclosporine-treated renal transplant recipients, in whom fish oils also induced a better response of glomerular filtration rate to amino acid infusion [12]. In a further placebo-controlled, randomized, double-blind study in stable cyclosporine-treated renal transplant recipients, ω3 PUFA caused a progressive increase in glomerular filtration rate and effective renal plasma flow, and a significant, though moderate, reduction in blood pressure, suggesting favorable modifications of renal function [13]. Currently, at least two multicenter trials are in progress to assess the ultimate clinical efficacy of this preventive approach.

Immune Chronic Renal Diseases

Studies in Lupus Nephritis
Glomerulonephritis accompanying systemic lupus erythematosus is one of the most serious and life-threatening complications of this disease. A number of studies have demonstrated an increased kidney production of TX, possibly deriving from platelet activation, and correlated with platelet hyperaggregability and the deposition of platelet antigens which are frequently seen in lupus nephritis. ω3 PUFA might have multiple points of attack, including a reduction of the accelerated atherosclerosis, which has emerged as an important cause of morbidity and mortality in this disease, and which could be related to the immunological mechanisms of the disease itself. Following previous promising animal studies, a study has been performed with ω3 PUFA in 12 subjects with lupus nephritis. Although affecting favorably a number of biological variables considered to have relevance in the mechanisms of the disease (a reduction in platelet aggregation, an improvement in red blood cell deformability, a decrease in whole blood viscosity, a decrease in triglycerides, an increase in HDL-cholesterol and a decreased production of leukotriene B_4 (LTB_4) by neutrophils), ω3 PUFA had no effect on immune complex or anti-DNA antibody titers, or albuminuria [14]. A prospective placebo-controlled clinical trial with a crossover design is now under way in this disease [15].

Studies in IgA Nephropathy

IgA nephropathy is one of the most common chronic glomerular diseases, with a progressive course of proteinuria, hypertension and glomerulosclerosis. Although earlier studies had suggested a benign course for this disease with unknown pathogenesis, long-term studies have rather suggested that a large proportion of these patients develops end-stage renal failure within a period of 10–20 years [16]. For this disease, in which no clearly effective treatment is available, favorable results were reported with the use of ω3 PUFA a few years ago [17]. Eicosapentaenoic acid (EPA) treatment for 1 year was associated with a stabilization of renal function in 10 patients compared to a control group. However, two subsequent prospective trials (of 24 and 9 months duration respectively) were not able to confirm such results [18, 19]. Because of such discrepancies a new prospective controlled study was planned and started, as a pilot phase, in 1987. Besides confirming, once more, favorable results on a number of investigated variables such as blood pressure, serum cholesterol and triglyceride levels, and platelet aggregation in response to epinephrine and ADP, this study showed a significant reduction in proteinuria and an improvement in renal function, as evaluated by iothalamate clearance [20]. These interesting results spurred in 1988 the start of a large multicenter trial on IgA nephropathy, conducted by the Mayo Nephrology Collaborative Group and involving 22 nephrological groups throughout the United States. This study is still in progress.

Studies of Renal Function in Other Chronic Glomerular Diseases

There are very few such clinical studies published in the literature, and all are short-term studies examining parameters of renal function in patients with chronic renal disease of various origin. In the first published study of this kind, 15 patients with chronic glomerulonephritis, chronic pyelonephritis, adult polycystic kidney disease, nephrosclerosis or tubulointerstitial nephritis were studied [21]. Patients were examined after 4 weeks of protein restriction, 4 weeks of high protein intake and after 8 weeks of dietary supplementation with ω3 PUFA together with a high protein intake. The highest values for effective renal plasma flow and glomerular filtration rate were observed after the last treatment period. Also, calculated renal resistance fell significantly, about 20% after ω3 PUFA supplementation. Interestingly, the rise in glomerular filtration rate during ω3 PUFA supplementation was not accompanied by a rise in proteinuria. Twenty-four-hour protein excretion was actually significantly reduced compared to the period of high protein intake [21]. In another study from the same group, a few other patients with chronic renal insufficiency from nonspecified origin were given ω3 PUFA supplementation for 12 weeks. In this study glomerular filtration rate did not change significantly, but a trend towards an increase in effective renal plasma flow was evident [22]. Results from these studies have been summarized in a recent review by the same group, concluding that a significant, although modest, increase in both glomerular filtration rate and effective renal plasma flow was induced by ω3 PUFA [23]. Our group has recently completed another trial in patients with chronic glomerular disease consisting of either membranous glomerulonephritis or idiopathic focal glomerular sclerosis [24]. None of our patients had renal insufficiency, but all had significant proteinuria without overt nephrotic syndrome. Besides the expected beneficial changes in serum triglycerides and serum generation of TX, a 6-week course of ω3 PUFA treatment with either a triglyceride or an ethyl ester preparation resulted in an increase in the ω3/ω6 PUFA ratio in plasma phospholipids, decreased proteinuria significantly, but did not affect creatinine or creatinine clearance [24]. Interestingly, but consistent with some prolonged effects of ω3 PUFA, the changes in proteinuria tended to persist for several (> 10) weeks after the end of treatment. Measurement of urinary renal metabolites of TX, prostacyclin and prostaglandin E_2 (PGE_2) in this study showed a similar reduction in all the dienoic prostanoids measured. It was not possible therefore to attribute the beneficial effect observed to a shift in the balance between vasoconstrictive (TX) and vasodilatory prostanoids (prostacyclin and PGE_2). On the basis of the encouraging results from this pilot study, a larger trial is currently under way.

Conclusions

ω3 PUFA have many attractive properties for use in kidney patients, including a very favorable benefit-risk profile. In humans, the most promising evidence for their use lies, for the moment, in protection from cyclosporine nephrotoxicity. However, other clinical applications are under study, including their use in lupus nephritis, IgA nephropathy and chronic glomerular disease. ω3 PUFA are to be considered at present a promising investigational approach to treatment of some renal diseases, but further clinical studies are needed to establish their efficacy and spectrum of clinical usefulness.

References

1   Hamazaki T, Nakazawa R, Tateno S, Shishido H, Isoda K, Hattori Y, Yoshida T, Fujita T, Yano S, Kumagai A: Effects of fish oil rich in eicosapentaenoic acid on serum lipid in hyperlipidemic hemodialysis patients. Kidney Int 1984;26:81–84.
2   Tenschert W, Rossodivita T, Rolf N, Winterberg B, Lison AE, Raidt H, Dorst KG, Zumkley H: Eignen sich Omega-3-Fettsäuren zur Therapie der Dyslipoproteinämie bei Hämodialysepatienten? Nieren Hochdruckkrankh 1988;17:106–110.

3   Rolf N, Tenschert W, Lison AE: Results of a long-term administration of omega-3 fatty acids in haemodialysis patients with dyslipoproteinaemia. Nephrol Dial Transplant 1990;5:797–801.

4   Rylance BP, Jordge MP, Saynor R, Parson V, Weston MJ: Fish oil modifies lipids and reduces platelet aggregability in hemodialysis patients. Nephron 1986;43:196–202.

5   Azar R, Dequiedt F, Awada J, Dequiedt P, Tacquet A: Effects of fish oil rich in polyunsaturated fatty acids on hyperlipidemia of hemodialysis patients. Kidney Int 1989;36(suppl 27):S239–S242.

6   van Acker BAC, Bilo HJG, Popp-Snijders C, van Bronswijk H, Oe PL, Donker AJM: The effect of fish oil on lipid profile and viscosity of erythrocyte suspensions in CAPD patients. Nephrol Dial Transplant 1987;2:557–561.

7   Lempert KD, Rogers JS II, Albrink MJ: Effects of dietary fish oil on serum lipids and blood coagulation in peritoneal dialysis patients. Am J Kidney Dis 1988;11:170–175.

8   Bakker DJ, Heberstroh BN, Philbrick DJ, Holub B: Triglyceride lowering in nephrotic syndrome patients consuming a fish oil concentrate. Nutr Res 1989;9:27–34.

9   Bennett WM, Elzinga L, Kelley V: Pathophysiology of cyclosporine nephrotoxicity: Role of eicosanoids. Transplant Proc 1988;20(suppl 3):628–633.

10 Sweny P, Wheeler DC, Lui SF, Amin NS, Barradas A, Jeremy Y, Mikhailidis DP, Varghese Z, Fernando ON, Moorhead JF: Dietary fish oil supplements preserve renal function in renal transplant recipients with chronic vascular rejection. Nephrol Dial Transplant 1989;4:1070–1075.

11 Stoof TJ, Korstanje MJ, Bilo HJ, Starink TM, Hulsmans RF, Donker AJ: Does fish oil protect renal function in cyclosporin-treated psoriasis patients? J Intern Med 1989;226:437–441.

12 Homan van der Heide JJ, Bilo HJG, Donker AMJ, Wilmink JM, Sluiter WJ, Tegzess AM: Dietary supplementation with fish oil modifies renal reserve filtration capacity in postoperative, cyclosporine A-treated renal transplant recipients. Transplant Int 1990;3:171–175.

13 Homan van der Heide JJ, Bilo HJG, Tegzess AM, Donker AJM: The effects of dietary supplementation with fish oil on renal function in cyclosporine-treated renal transplant recipients. Transplantation 1990;49:523–527.

14 Clark WF, Parbtani A, Huff MW, Reid B, Holub BJ, Falardeau P: Omega-3 fatty acid dietary supplementation in systemic lupus erythematosus. Kidney Int 1989;36:653–660.

15 Donadio JV Jr: Omega-3 polyunsaturated fatty acids: A potential new treatment of immune renal disease. Mayo Clin Proc 1991;66:1018–1028.

16 D'Amico G, Imbasciati E, Belgioioso GDI, et al.: IgA mesangial nephropathy. Clinical and histological study of 374 patients. Medicine 1985;64:49–60.

17 Hamazaki T, Tateno S, Shishido H: Eicosapentaenoic acid and IgA nephropathy. Lancet 1984;ii: 1017–1018.

18 Bennett WM, Walker RG, Kincaid-Smith P: Treatment of IgA nephropathy with eicosapentaenoic acid: A two-year prospective trial. Clin Nephrol 1989;31:128–131.

19 Cheng IKP, Chan PCK, Chan MK: The effect of fish-oil dietary supplement on the progression of mesangial IgA glomerulonephritis. Nephrol Dial Transplant 1990;5:241–246.

20 Donadio JV Jr, Holman RT, Holub BJ, Bergstralh EJ: Effects of omega-3 polyunsaturated fatty acids in mesangial IgA nephropathy (abstract). Kidney Int 1990;3:255.

21 Schaap GH, Bilo HJG, Popp-Snijders C, Oe PL, Mulder C, Donker AJM: Effects of protein intake variation and omega-3 polyunsaturated fatty acids on renal function in chronic renal disease. Life Sci 1987;41:2759–2765.

22 Schaap GH, Bilo HJG, Beukhof JR, Gans ROB, Popp-Snijders C, Donker AJM: The effect of short-term omega-3 polyunsaturated fatty acid supplementation in patients with chronic renal insufficiency. Curr Ther Res 1991;49:1061–1070.

23 Bilo HJG, Homan van der Heide JJ, Gans Rob, Donker AJM: Omega-3 polyunsaturated fatty acids in chronic renal insufficiency. Nephron 1991;57:385–393.

24 De Caterina R, Caprioli R, Giannessi D, Sicari R, Galli C, Lazzerini G, Bernini W, Carr L, Rindi P: n-3 polyunsaturated fatty acids reduce proteinuria in patients with chronic glomerular disease. Kidney Int 1993;44:843–850.

Raffaele De Caterina, MD, Vascular Medicine and Atherosclerosis Unit,
Cardiovascular Division, Department of Medicine, Brigham and Women's Hospital,
Harvard Medical School, 221 Longwood Avenue, Boston, MA 02115 (USA)

Galli C, Simopoulos AP, Tremoli E (eds): Effects of Fatty Acids and Lipids in
Health and Disease. World Rev Nutr Diet. Basel, Karger, 1994, vol 76, pp 143–145

Placebo-Controlled Trials of ω3 Fatty Acids in Chronic Inflammatory Bowel Disease[1]

R. Lorenz, K. Loeschke

Institut für Prophylaxe der Kreislaufkrankheiten und Medizinische Klinik,
Klinikum Innenstadt, Universität München, Germany

A potential benefit from ω3 fatty acids in Crohn's disease (CD) and ulcerative colitis (UC) is suggested by several lines of evidence: (1) increased levels and synthetic capacity, which correlate to disease activity, support a pathogenic role of arachidonic acid (AA)-derived proinflammatory mediators and cytokines in both forms of chronic inflammatory bowel disease (CIBD); (2) the therapeutic mainstays in CIBD, 5-aminosalicylic acid compounds and steroids, have been shown in vitro or in vivo to suppress the formation of leukotrienes; (3) ω3 fatty acids interfere with AA-derived eicosanoid formation and action as well as cytokine synthesis, and (4) the epidemiologic evidence and studies in animal models are compatible with a beneficial effect of dietary ω3 fatty acids in CIBD.

Therefore we conducted the first prospective, randomized, double-blind, placebo-controlled crossover trial of 3.2 g/day ω3 fatty acid supplementation in 39 patients with CIBD of low to moderate activity [1]. At the expense of AA, ω3 fatty acids were incorporated into mucosa even better than into plasma phospholipids and eicosanoid production was blunted. Clinical disease activity was improved just short of significance in patients with UC but not with CD and macroscopic mucosal inflammation appeared moderately reduced in both forms of CIBD.

Since then, there have been at least four more placebo-controlled trials. The structure and main findings of the trials are listed in table 1. In a multicenter crossover trial coordinated in St. Louis, Mo., USA [2], 24 patients

[1] Supported by Bundesministerium für Forschung und Technologie (07ERG03/7).

Table 1. Placebo-controlled trials of ω3 fatty acids in chronic inflammatory bowel disease

Ref.	Structure of trials						Major results					
	activity at entry	n	conventional therapy	ω3 g/day	duration months	washout months	clinical activity	time to relapse	steroid sparing	inflammation macroscopic	inflammation histologic	mediator synthesis
Ulcerative colitis												
1	+/++	10	minimized and continued	3.2	2 × 3*	1	(+)			(+)		+
2	++	24	5-ASA continued steroids adapted	5.4	2 × 4*	1	(+)		(+)	(+)	+	+
3	+/++	11	adapted	4.2	2 × 3*	2	+		(+)		±	±
4	+	40	continued	5.6	12		(+)	±	+			+
	+++	56	steroids adapted									
Crohn's disease												
1	+/++	29	minimized and continued	3.2	2 × 3*	1	±			(+)		+
5	+/++	204	withdrawn after 8 weeks	5	12		?	±		?	?	

± = Unchanged; (+) = suggestive trend; + = significant beneficial effect; ? = not yet reported; * = crossover design.

with active UC were allocated to 5.4 g/day ω3 fatty acids or vegetable oil for 4 months each, given on top of conventional therapy. ω3 fatty acids significantly improved rectal LTB$_4$ levels, histologic scores of mucosal inflammation and body weight but not other symptoms and signs. They also tended to improve macroscopic inflammation and steroid requirement.

In a crossover study from California [3], 11 patients with mild to moderate UC were treated with 4.2 g/day ω3 fatty acids or olive oil for 3 months each as adjunct to drug therapy. Significant improvement of clinical disease activity was reported and drug requirement was lowered, without major changes in mucosal LTB$_4$ levels and histologic activity of inflammation.

In the largest trial in UC so far, conducted in England [4], 56 patients in relapse and 40 patients in remission were treated with 5.6 g/day ω3 fatty acids or olive oil for 1 year added to drug therapy. ω3 fatty acids accelerated remission and helped sparing steroids after relapse, but relapse rate was not reduced. There is only a preliminary report of a multicenter trial with 5 g/day ω3 fatty acids for 1 year compared to olive oil or carbohydrate-poor diet in 204 patients with CD [5]. After remission induction with steroids, relapse rate in CD was also not reduced by ω3 fatty acids.

In summary, current evidence from controlled trials does not yet allow definite conclusions to be drawn. In UC, a limited clinical benefit appears likely with ω3 fatty acid supplements given on top of drug therapy from consistent positive trends in several trials. ω3 fatty acids appear to accelerate remission induction, spare steroids and moderately reduce activity in UC in remission. However, a major impact of ω3 fatty acids during the first year of supplementation on cumulative incidence of relapse is unlikely both in UC and CD.

References

1   Lorenz R, Weber PC, Szimnau P, et al: Supplementation with n-3 fatty acids from fish oil in chronic inflammatory bowel disease: A randomized placebo-controlled, double-blind cross-over trial. J Intern Med 1989;225(suppl 1):225–232.
2   Stenson WF, Cort D, Rodgers J, et al: Dietary supplementation with fish oil in ulcerative colitis. Ann J Med 1992;116:609–614.
3   Aslan A, Triadafilopoulos G: Fish oil fatty acid supplementation in active ulcerative colitis: A double-blind, placebo-controlled crossover study. Am J Gastroenterol 1992;87:432–437.
4   Hawthorne AB, Daneshmend TK, Hawkey CJ, et al: Treatment of ulcerative colitis with fish oil supplementation: A prospective 12-month randomized controlled trial. Gut 1992;33:922–928.
5   Lorenz-Meyer H, Purrmann J, Scheurlen C, et al: Crohnstudie V: Ergebnisse der Studie zur Erhaltung der Remission bei Morbus Crohn mit ω-3 FS bzw. einer kohlenhydratarmen Kost (abstract). Z Gastroenterol 1992;30:654.

R. Lorenz, MD, Universität München, Pettenkoferstrasse 9, D–80336 München (Germany)

Galli C, Simopoulos AP, Tremoli E (eds): Effects of Fatty Acids and Lipids in
Health and Disease. World Rev Nutr Diet. Basel, Karger, 1994, vol 76, pp 146–148

....

ω3 Fatty Acids and Respiratory Disease

Howard R. Knapp

Division of Clinical Pharmacology, Department of Internal Medicine,
University of Iowa, Iowa City, Iowa, USA

Although epidemiological data has suggested a lower incidence of respiratory diseases among Greenland Eskimos [1], it is apparent that genetics and
environmental influences other than diet play a significant role. In contrast to
the likely cardiovascular benefits, there is little population data associating ω3
fatty acid ingestion with the presence or absence of specific respiratory disorders. Because leukotrienes have a significant pathogenic role in allergic and
pulmonary disorders, the observation that dietary ω3 fatty acids influence the
types and amounts of leukotrienes synthesized has prompted studies of their
effect in clinical respiratory disorders. This paper will briefly review published
studies in the area of ω3 fatty acids and the human respiratory tract, with
discussion of animal studies only where they are pertinent to clinical investigations.

Studies in Human Asthma

There have been at least five studies published on the effects of ω3 fatty
acid ingestion in human asthma. Payan et al. [2] found that ingestion of 4 g/day
of pure eicosapentaenoic acid (EPA) ethyl ester for 8 weeks caused decreased
neutrophil chemotaxis and leukotriene production ex vivo, but had no effect on
clinical parameters or medication use in 12 asthmatics, 3 of whom had had
documented adverse responses to aspirin. Arm et al. [3] also studied dietary ω3
fatty acids over a 6-week period in allergic asthmatics, and found no change in
symptom scores, bronchodilator use, or morning peak expiratory flow rates. In
a subsesquent report [4], they performed allergen challenge testing in their

patients and while no effect on the immediate bronchoconstriction was found, there appeared to be a lessening of the late-phase response in the group on fish oil versus those taking placebo. One problem of interpretation, however, rests on the much greater late-phase pretreatment response in the fish oil group than in those on placebo, so that the response on placebo is only slightly greater than that seen after fish oil, i.e. the fish oil effect was seen only when compared to baseline and not to placebo.

The report of Dry and Vincent [5] suggested a statistical benefit for asthmatics ingesting ω3 fatty acids after 9 months, but not before or after this time point. Their 12 patients were divided into two 'randomized' groups that had baseline forced expiratory volume at 1 min (FEV_1s) of 60% (control) and 73% (ω3 fatty acids) of predicted. Although baseline pulmonary function tests (PFTs) were not statistically different in these small groups of patients, it is clear that the groups were not well matched and unsurprising that they might diverge 'significantly' at one of the five time points. Overall, then, the studies carried out to date have not suggested a sufficient benefit to warrant a major research effort in asthma, despite interesting ex vivo biochemical findings. The work of Payan et al. [2] mentioned above did not find that three aspirin-sensitive asthmatics experienced any decrement of pulmonary function attributable to ω3 fatty acids, but a later report by Picado et al. [6] indicated that 3 g EPA/day caused a persistent worsening of peak expiratory flow in 10 aspirin-sensitive asthmatics. This effect did not appear until 5 weeks of supplementation, and was not associated with worsening self-reported symptom scores. Interestingly, the authors suggested that EPA could be acting as a cyclooxygenase inhibitor in their patients. Aspirin-induced bronchospasm, however, is notable for the rapid development of tachyphylaxis, so that most subjects taking daily aspirin will become desensitized to its effects in 3 days or so. Exactly how this might relate to an in vivo effect of ω3 fatty acids in these patients is unclear.

Studies in Allergy

Ingesting ω3 fatty acids does not result in a clinical improvement of asthma, but worsening of immediate-hypersensitivity reactions was found in studies showing increased synthesis of leukotrienes and unchanged or worsened symptoms in guinea pig systemic anaphylaxis [7], cardiac anaphylaxis [8], mouse footpad anaphylaxis [9], rat cutaneous allergy [10], or human nasal allergen challenge [11]. Total cysteinyl-leukotriene production ex vivo was unchanged in human subjects taking ω3 fatty acids, although urinary LTE_4 was reduced 35% [12]. No description of urinary LTE_5 synthesis was provided, however, and since the 4- and 5-series leukotrienes are considered to be equipotent, it is not known whether total leukotriene synthesis was altered.

Although ω3 fatty acids do not appear to be therapeutic in allergic disorders or asthma, observations on their bactericidal effects [13] and in a rat model of hypobaric hypoxia [14] suggest that studies in chronic lung disorders may prove interesting. The former indicates a possible mechanism whereby chronic lung infections become colonized by pseudomonas and enterobacteriaciae, while in the latter study survival, right ventricular hypertrophy and pulmonary pressures of hypoxemic rats were much improved by dietary ω3 fatty acids [14]. Only further work in patients with chronic lung disease, for whom no therapy exists in many cases, will define the therapeutic utility of dietary ω3 fatty acids.

References

1   Kromann N, Green A: Epidemiological studies in the Upernavik District, Greenland. Acta Med Scand 1980;208:401–406.
2   Payan DG, Wong MY, Chernov-Rogan T, et al: Alterations in human leukocyte function induced by ingestion of eicosapentaenoic acid. J Clin Immunol 1986;6:402–410.
3   Arm JP, Horton CE, Mencia-Huerta JM, et al: Effect of dietary supplementation with fish oil on mild asthma. Thorax 1988;43:82–92.
4   Arm JP, Horton CE, Spur BW, et al: The effects of dietary supplementation with fish oil on the airways response to inhaled allergen in bronchial asthma. Am Rev Resp Dis 1989;139:1395–1400.
5   Dry J, Vincent D: Effect of a fish oil diet on asthma: Results of a 1-year double-blind study. Int Arch Allergy Appl Immunol 1991;95:156–157.
6   Picado C, Castillow JA, Schinca N, et al: Effects of a fish oil-enriched diet on aspirin-intolerant asthmatic patients: A pilot study. Thorax 1988;43:93–97.
7   Lee TH, Israel E, Drazen JM, et al: Enhancement of plasma levels of biologically active leukotriene B compounds during anaphylaxis in guinea pigs pretreated by indomethacin or by a fish oil-enriched diet. J Immunol 1986;136:2575–2582.
8   Juan H, Peskar BA, Simmet T: Effect of exogenous eicosapentaenoic acid on cardiac anaphylaxis. Br J Pharmacol 1987;90:315–325.
9   Yoshino S, Ellis EF: Stimulation of anaphylaxis in the mouse footpad by dietary fish oil fatty acids. Prostagaldins Leukot Essent Fatty Acids 1989;36:165–170.
10 Prickett JD, Robinson DR, Bloch KJ: Enhanced production of IgE and IgG antibodies associated with a diet enriched in eicosapentaenoic acid. J Immunol 1982;46:819–826.
11 Knapp HR: In vivo synthesis of leukotriene B_5 during nasal allergen challenge of fish oil-supplemented subjects. Clin Res 1991;39:319A.
12 Von Schacky C, Kiefel R, Jendraschak E, et al: n-3 fatty acids and cysteinyl-leukotriene formation in humans in vitro, ex vivo, and in vivo. J Lab Clin Med 1993;121:302–309.
13 Knapp HR, Melly MA: Bactericidal effects of polyunsaturated fatty acids. J Infect Dis 1986;154:84–94.
14 Archer SL, Johnson GJ, Gebhard RL, et al: Effect of dietary fish oil on lung lipid profile and hypoxic pulmonary hypertension. J Appl Physiol 1989;66:1662–1673.

Howard R. Knapp, MD, PhD, Division of Clinical Pharmacology, Department of Internal Medicine, C-31 GH, University of Iowa, Iowa City, IA 52242 (USA)

Subject Index

Crohn's disease, effect of dietary ω3 fatty
    acids 143–145
Cyclooxygenase, preeclampsia role 115, 116
Cyclosporin, dietary ω3 fatty acid effect on
    nephrotoxicity 138, 139

Δ6-Desaturase
    defects in cancer 77
    nutritional effects on activity 107
Diabetes
    cardiovascular complications 16–18, 51
    dietary management
        type I 15, 16
        type II 15, 17, 18
    effect of dietary ω3 fatty acid
        supplementation 15–18
Docosahexaenoic acid (22:6ω3)
    antithrombotic activity 3, 4
    effect
        cancer therapy, susceptibility of
            tumor 82, 84
        hypertension 53
        LDL metabolism 30–33
        platelet aggregation 46, 48
        VLDL metabolism 26–29
    neonate status 119, 120
    protection
        carcinogenesis 64, 66–68
        coronary heart disease 1–7
    role
        fetal development 105–107
        preeclampsia 103, 111, 112
    sources 38
Doxorubicin, ω3 fatty acid effect on cancer
    therapy 82, 84

Eicosanoids, effect of dietary ω3 fatty
    acids 4, 10, 60, 61, 95, 99
Eicosapentaenoic acid (20:5ω3)
    antithrombotic activity 3, 4
    cancer cell susceptibility 77, 78
    effect
        cachexia 86–88
        LDL metabolism 30–33
        platelet aggregation 48
        tumor growth 86–88
        VLDL metabolism 26–29
    protection
        carcinogenesis 64, 66–68
        coronary heart disease 1–7

sources 38
status in pregnancy 127–129
Endothelial-derived relaxation factor, effect
    of dietary ω3 fatty acids 4, 48, 49
Eskimos, incidence of disease
    cancer 66
    coronary heart disease 1, 47, 52, 130,
        133–135
    respiratory disease 146
Essential fatty acids
    function 122
    types 122

Glomerular filtration rate, effect of dietary
    ω3 fatty acids 140, 141
Glomerulonephritis, see Autoimmune
    glomerulonephritis

High-density lipoprotein
    effects of ω3 fatty acids
        diabetes 16, 17
        metabolism 34–37, 41, 42, 44
        subfraction distribution 35–37
    synthesis 31, 32
H-ras oncogene, suppression by ω3 fatty
    acids 67
Hypercholesterolemia, see Cholesterol
Hyperlipidemia
    dietary management 38–40, 52, 137,
        138
    renal patients 137, 138
    risk factor for coronary heart disease 39
Hypertension, see Blood pressure,
    Preeclampsia
Hyperthermia, ω3 fatty acid effect on
    cancer therapy 82

Immunoglobin A nephropathy, effect of
    dietary ω3 fatty acids 140
Interleukin-1
    biological function 89, 90
    effect of dietary ω3 fatty acids 46,
        90–92, 99
    fatty acid suppression mechanism 99
    synthesis modulators 90, 91
Intrauterine growth retardation
    complications 105, 106
    essential fatty acid status 105–108
    fetal morbidity 105
Isoprenoids, see Prenylation

Kidney, effect of dietary ω3 fatty acids 131, 137–141

Lecithin cholesterol acyl transferase, effect of dietary ω3 fatty acid on activity 36, 37
Leukemia, effect of dietary ω3 fatty acids 64
Leukocyte, dietary ω3 fatty acid effect on function 49
Leukotriene B$_4$
  effect of dietary ω3 fatty acids 4, 10, 60, 61, 95, 99, 145
  role in respiratory disease 146, 148
Linoleic acid (18:2ω6)
  cholesterol-lowering activity 41, 42
  effects
    carcinogenesis 66
    lipid metabolism 41, 42
  levels in preeclampsia 110
  sources 38
α-Linolenic acid (18:3ω3)
  effect on lipoprotein metabolism 43
  sources 38
γ-Linolenic acid
  cancer cell susceptibility 77–79
  pancreatic cancer therapy 74–76
Low-density lipoprotein
  effects of dietary fatty acids 2, 16, 30–33, 39–44, 52
  metabolism 30 31
  receptor density 32, 42
  size 39, 40
Lupus nephritis, effect of dietary ω3 fatty acids 139

Melanoma, effect of dietary ω3 fatty acids 64
Monocyte
  dietary ω3 fatty acid effect on function 46, 55–58
  tissue factor, *see* Tissue factor

Neonate, essential fatty acid status effect on long-term development 104, 119–121
Neutrophil, ω3 fatty acid effect on chemotaxis 99

Oleic acid (18:1ω9), sources 38
Oxidation, fatty acid, effect of dietary ω3 fatty acids 28

Pancreatic cancer
  effect of dietary ω3 fatty acids 74–76
  γ-linolenic acid therapy 74–76
Parinaric acid, tumor cytotoxicity 83
Phospholipids
  cancer role 68
  effect of dietary ω3 fatty acids 68
Platelet aggregation
  effect
    arachidonic acid 60–62
    ω3 fatty acids 4, 45–48, 53, 60–62
  mechanism 62
Preeclampsia
  antioxidant status 114–116
  aspirin effects 116
  complication of intrauterine growth retardation 105, 106
  docosahexaenoic acid levels 103, 111, 112
  essential fatty acid status 110–112, 117–119
  incidence 110
  lipid peroxidation 103, 114–116
  phospholipid composition 111, 112
  role
    arachidonic acid 127–129
    cyclooxygenase 115, 116
    eicosapentaenoic acid 127–129
    prostacyclin 107, 108, 110, 116, 126
    thromboxane 107, 108, 110, 115, 116, 126
Pregnancy
  antioxidant status 114, 116
  fatty acid status 127–129
  prostaglandin status 123
Pregnancy-induced hypertension, *see* Preeclampsia
Prenylation
  biological function 70, 71
  enzyme catalysis 70
  inhibition 72, 73
  neoplasia relationship 64, 65
  oncogenesis role 71, 72
  ras proteins 71, 72
Promotion, cancer, protective effects of ω3 fatty acids 67, 68